POWER STRUCTURE

Ownership, Integration, and Competition in the U.S. Electricity Industry

POWER STRUCTURE

Ownership, Integration, and Competition in the U.S. Electricity Industry

by

John E. Kwoka, Jr.
George Washington University

KLUWER ACADEMIC PUBLISHERS
Boston/Dordrecht/London

Distributors for North America:
Kluwer Academic Publishers
101 Philip Drive
Assinippi Park
Norwell, Massachusetts 02061 USA

Distributors for all other countries:
Kluwer Academic Publishers Group
Distribution Centre
Post Office Box 322
3300 AH Dordrecht, THE NETHERLANDS

Library of Congress Cataloging-in-Publication Data
Kwoka, John E.
    Power structure : ownership, integration, and competition in the
U.S. electricity industry / by John E. Kwoka, Jr.
        p.   cm.
    Includes bibliographical references and index.
    ISBN 0-7923-9843-2
    1. Electric utilities--United States.   2. Electric utilities-
-Government ownership--United States.   3. Privatization--United
States.   I. Title.
    HD9685.U5K87   1996
    333.793'2'0973--dc21   96-47148
                                                                            CIP

Printed on acid-free paper.

Printed in the United States of America

Dedication

In honor of my father, John Kwoka, and in memory of my mother, Mary Kwoka.

Contents

PREFACE

Most books are long journeys. The first step for this project took place in the fall of 1990 when I was asked about my possible interest in researching the relative performance of publicly owned versus privately owned electric utilities in the U.S. The question was intriguing for several reasons. In contrast to conventional economic wisdom that predicts lower costs and prices under *private* ownership, many if not most studies of electric power show that *public* ownership results in superior performance. The reasons for this apparent anomaly have long been disputed. Understanding them would seem to be increasingly important in an era when many countries have been undertaking, or at least contemplating, privatization of their electricity industries.

In addition, the U.S. electric power industry has been poised on the threshold of its own fundamental restructuring. Here the principle issues are not ownership but, rather, vertical deintegration of generation from transmission and distribution, and the potential role of competition. Utilities in this country, however, already exhibit considerable diversity in their degree of integration and of competition. That diversity could be used in the research to gain at least some insight into the effects of the more sweeping changes under consideration. And looking ahead, a further useful outcome of this project would be to establish benchmarks of performance for evaluating a host of such policy initiatives.

Cinching the case was every economists' irresistible lure — new data. Data had recently become available covering many more publicly owned systems on a basis comparable privately owned utilities. This opened up the possibility of substantially more definitive research on all of these issues than previously possible.

I therefore soon agreed to undertake the project. What I did not appreciate at the time was the full extent of the odyssey on which I was embarking. Compilation of the necessary data proceeded (off and on, to be sure) for nearly four years and was not entirely complete even then. Another year was devoted to obtaining the core analytical results. Writing the present monograph consumed an additional year and a half, although throughout there have been the usual interruptions from the academic schedule and other commitments.

The outcome of the project has nonetheless been very satisfying. The substantive results cast light on key determinants of the cost and price performance of U.S. electric utilities: vertical integration versus deintegration, public versus private ownership of enterprises, and retail competition versus monopoly distribution. Many other questions are addressed as well. These include the effects of power pools, holding companies, incentive regulation, combination gas and electric utilities, utility commission characteristics, and differences among customer classes.

These questions are matters of longstanding interest for the U.S. electricity industry. In addition, they have direct relevance for the restructuring of the electric power sectors both here and in other countries. It would be presumptuous to suppose that any single piece of research could definitively answer such questions, of course, and this research no doubt has its limitations. But it is not unrealistic to hope that its conclusions will contribute to a deeper understanding of these issues and at the same time advance the policy debate on electric power reform. To the extent that it does, this book will have succeeded in its objectives.

This project has benefitted crucially from assistance from numerous individuals and institutions. For their comments on various interim results, I wish to thank seminar participants at the George Washington University Economics Department, the 1994 American Economics Association session "Regulation and Competition," the Harvard University Economics Department and Kennedy School, MIT Sloan School, the Antitrust Division of the Justice Department, the Brookings Institution, Resources for the Future, the Williamsburg Conference of the Institute for Public Utilities, the American University Department of Economics, and the Department of Justice/Federal Trade Commission Electricity Workshop.

I also owe a debt of gratitude to a number of individuals for helpful discussions, suggestions, information, or manuscript-reading. These include Paul Joskow, John Mayo, Michael Pollitt, Robin Prager, Steve St. Marie, Chris Snyder, and Larry White. Assistance in data compilation and analysis and in manuscript preparation was ably provided by Kevin Roth and Theresa Alafita. I owe special thanks to Aimee Dimmerman for bringing the pieces of the manuscript together at the very end. Some research assistance was supported through the Center for Economic Research at GWU.

Much of the actual writing of this book was done while I was on sabbatical from George Washington University, first as Visiting Professor of Economics at Harvard and then as Guest Scholar at Brookings. I owe Dick Caves, Bob Crandall, and Henry Aaron my thanks for the opportunities to reside in the productive environments of their institutions. I also wish to thank Joe Cordes, my department chair at George Washington University, for his unwavering support throughout this project, and Zac Rolnik, my editor at Kluwer, for his enthusiastic and efficient stewardship.

My debt could not be greater to John Kelly and the American Public Power Association. At Lorel Wisniewski's suggestion, it was John who, as Director for Economics and Research at APPA, made the initial inquiry as to my interest in the project. The APPA subsequently provided financial assistance for project formulation, for my initial research into the ownership issue, and for some follow-up activities. They also undertook the truly enormous tasks of data compilation and reconciliation. Kaylyn Hipps, Diane Moody, and Julie Murphy deserve special thanks for their efforts, as does JoAnn Hutchison for facilitating countless matters between myself and APPA. Throughout the process John Kelly contributed most of all, patiently responding to an endless stream of questions and requests for additional data, while insisting on the highest standards of research objectivity regardless of the conclusions that might emerge from this work.

I should note, of course, that none of these individuals or institutions is responsible for any errors in this volume, nor do they necessarily endorse any of its conclusions.

Finally, I would like to acknowledge those who in so many other ways have helped on my longer personal and intellectual journey to this point. My parents who years ago gave so much for my education; teachers, colleagues, and students who subsequently have taught me industrial organization economics; and my wife and daughter for their continuing support and patience deserve recognition. In a very real sense this book is the product of their efforts, too, and I am grateful to all.

Chapter 1

INTRODUCTION

The electric power industries in the United States and most other countries have long been structured in accordance with a simple paradigm. In the belief that electric power is a natural monopoly where a single company can provide service most efficiently, this paradigm involves large franchised utilities with exclusive service territories and which are integrated into generation, transmission, and distribution. Concerns over price and cost distortions from monopoly power are addressed by regulation if the utilities are privately owned, or by outright public ownership.

Over the past two decades, however, this conventional paradigm has been subject to fundamental challenge. Numerous countries have grown dissatisfied with the performance of their large, integrated, monopoly electricity sectors, whether state-owned or regulated. Some have adopted thorough-going policies of privatization, deintegration, deregulation, and direct competition among sellers. Other countries have sought to reduce, if not eliminate, the role of government in the decisions and the operations of industry. These policies have represented an historic alteration both in the governance of enterprises and in the role of government itself.

These changes have already had discernible effects in the electric power industry, but in reality they represent only the beginnings of a sweeping restructuring. Over the next decade many more countries will adopt alternative means of achieving the objectives of efficiency in electricity production and prices that reflect minimum economic costs. While change is certain, the diversity of approaches currently employed and being considered makes it difficult to predict the final structure of the transformed industry.

In some ways, the United States electric power industry corresponds to the traditional paradigm. The largest and most familiar segment of this industry consists of about 200 privately owned companies. These "investor-owned utilities" (IOUs) are individually very large and vertically integrated, typically represent monopolies within their service territories, and are subject to both state and federal regulation. Their structure results from the paradigmatic belief in the importance of economies of scale; economies from coordinating generation, transmission, and distribution within a single company; private ownership for overall cost efficiency, and regulation as a mechanism for overseeing price and investment decisions.

Yet the U.S. industry differs from that of virtually all other countries in the diversity of firm structures that coexist. In addition to these IOUs, there are fully 2,000 publicly owned electric utilities, 1,000 rural electrical cooperatives, and a handful of federal power projects. Both the publicly owned systems and rural coops tend to be much smaller in size and are often (but not always) *un*integrated. That is, rather than generating their own power, they purchase much or all of their requirements from IOUs and federal projects and then distribute power to customers within their service territories. The federal power projects, by contrast, are almost exclusively generators for the wholesale market. In addition to this diversity in terms of ownership and integration, it should be noted that the U.S. industry includes a modest number of "competitive" utilities. In these cases communities are served by two different utilities, which may even have duplicate facilities and allow customers to switch suppliers.

The diversity of the U.S. electric power industry provides an opportunity to examine the performance implications of structure and governance differences among firms and thereby to anticipate the consequences of further imminent changes in the industry. More specifically, this study examines whether cost efficiencies are sacrificed by deintegration. If so, can these efficiencies be recaptured by power pools, holding companies, or other institutional adaptations? How important are scale economies in generation or distribution, either by themselves or in conjunction with any vertical economies? Does public ownership breed inefficiency, or does it actually have advantages in certain tasks, at least in comparison with regulation? Do prices differ depending upon such factors as ownership, distribution competition, or even method of selection of state utility commissioners, holding costs constant? Do these effects differ among residential, commercial, and industrial customers? If so, does this help to explain the origins of public ownership itself?

The further reason for examining the U.S. electric power industry revolves around current policy initiatives. Recent legislative and regulatory actions and proposals have set out a framework for transforming the electric power industry. The objectives of this transformation are precisely those noted at the outset — namely, efficiency in production and in pricing. The strategies being pursued involve at least partial deintegration of generation from distribution, regulatory reform or deregulation, and substantially greater competition in retail supply as well as generation. The degree to which these strategies may succeed can be illuminated by careful analysis of *existing* differences among utilities in these very dimensions of their structure and operations.

The long-standing nature of such differences makes the U.S. experience virtually unique. Individual utilities have remained public or private, integrated or deintegrated, competitive or monopoly for considerable periods of time, so that the enterprises are far more likely to be in equilibrium rather than in transition. By contrast, comparisons of performance of various enterprises in other countries are often confounded by the recent nature of their transformations to private, deintegrated, or competitive status. Moreover, many other experiences involve non-monopoly firms in traditional industries (for example, state-owned steel companies or airlines) where some form of competition may accompany privatization. Under these circumstances, attributing effects to privatization versus competition becomes considerably more difficult.

That said, a somewhat different challenge of determining causation arises in the case of the U.S. electric power industry. Its great diversity requires considerable analytical care to ensure that the effects of any one structural factor (vertical integration, or competition, for example) are distinguished from all others. Of course, many of the other causal factors — holding companies, tax effects, hydro power access, among others — are important in their own right and deserve attention. A further concern is that "private" utilities in the U.S. have traditionally been subject to cost-of-service regulation with its own well-documented behavioral and performance distortions. The result of this is that comparisons between publicly owned and "private" utilities may in fact involve comparing the inefficiencies of standard regulation with the inefficiencies of public ownership. On the other hand, for infrastructure industries such as electric power, this may be the *relevant* comparison, since privatization is typically accompanied by some form of social oversight and control such as regulation.

For these reasons, the U.S. electric power industry affords an unusual opportunity to assess the structural and governance causes of cost and price performance. This is the fundamental objective of the present study. In order to set the stage for the remainder of the analysis, we begin with a brief historical introduction to the U.S. electric power industry.

1.1 Historical Perspective on the U.S. Electric Power Industry

The origins of the U.S. electric power industry are usually traced to the September 1882 date on which Thomas Edison opened his Pearl Street Station in New York City. Despite its small size and mixed commercial success, this pioneering facility was a technological marvel (Hughes, 1983). The first customers were large business establishments, such as brokerage houses, soon followed by the city itself, which wanted power for street lighting purposes. Residential uses were initially a minor factor.

Cities and towns everywhere quickly sought to encourage electric power production within their jurisdictions. Some did so by granting franchises to private entrepreneurs, leading to some 151 private utilities by 1885 and over 2,500 by 1900 (see Table 1.1). In larger cities, multiple franchises were common, often nonexclusive, and sometimes explicitly duplicative ("competitive") with each other. Chicago, for example, awarded 45 franchises between 1882 and 1905, only one (for a small area) exclusive, 16 duplicative, and three covering the entire city (Jarrell, 1986, p. 292).

In smaller jurisdictions, private power initiatives were slower and less certain to materialize. Rather than waiting, many municipalities simply proceeded to build their own electric power generation plants. Power was then sold to local businesses as well as utilized for street lighting. From 15 public systems in 1885, the number grew to over 700 by the turn of the century.

Nonexclusive, even duplicative, franchises pitted private electric companies against each other in the competition for customers. The aggressive nature of such competition together with the high capital costs of electric power production sometimes resulted in prices covering only short-run operating costs, a circumstance that threatened companies' viability. On the other hand, where competition did *not* exist, profitability often rose to the point where privately owned utilities began to fear takeover by municipalities. For both these reasons, private utilities felt their future was quite uncertain.

A 1907 report of the National Civic Association on the relative merits of public versus private ownership criticized the performance of municipal utilities and concluded that regulation by expert *state* commissions might well be the best way to control natural monopolies (Anderson, 1981, pp. 44 – 47; Geddes, 1992). It even suggested that privately owned firms might escape public takeover by actively embracing such regulation. Beginning with Wisconsin and New York in 1907, regulatory commissions were established in two-thirds of the states by 1914 and in virtually all by the 1920s.

The premise of rate-of-return regulation was and is that rates should cover both operating costs and capital or financing costs, effectively establishing rates at something like average total cost. How effective such regulation is in actually achieving this objective — as opposed to just ratifying the exercise of monopoly power — has been the subject of debate ever since. This early period also saw the beginning of a vigorous debate between advocates of public versus private ownership. Municipally owned utilities portrayed private systems as unconstrained monopolists and cast themselves as defenders of consumers' access to power at "fair" prices. Privately owned systems charged municipal utilities with unfair competition due to the deep pockets of their public "owners" and argued that only they met the market test. Although there was some attention to the question of relative price and cost (Hausman and Neufeld, 1994), much of the argument of the

YEAR	PRIVATE	PUBLIC	COOPS	TOTAL
	Table 1.1. NUMBER OF ELECTRIC UTILITIES, BY OWNERSHIP SELECTED YEARS			
1885	151	15		166
1890	872	107		979
1895	1690	355		1945
1900	2514	732		3246
1905	3076	1109		4185
1912	3659	1737		5396
1917	4224	2411		6635
1922	3774	3014		6788
1927	2137	2320		4457
1932	1627	1863	0	3490
1937	1401	1921	126	3448
1942	NA	2078	803	NA
1947	858	2049	911	3818
1954	540	2019	1024	3583
1959	467	1990	1132	3489
1965	296	2034	939	3269
1970	299	2010	923	3232
1975	256	2224	982	3462
1980	217	2199	924	3340
1985	273	1966	958	3197
1990	267	2011	953	3231

Note: Total does not include federal power projects.
Sources: 1) Public Power, September/October 1982, p.68.
      2) DOE, Financial Statistics of Selected Electric Utilities, various years.

time was polemical in nature, heavy with overtones of the controversy about populism and socialism that dominated the era.

From the turn of the century into the 1920s, electric utilities proliferated to meet burgeoning demand for power. There were 4,224 privately owned systems[1] in 1917, but subsequent consolidation cut that number in half within ten years. Much of the consolidation was the result of acquisitions by holding companies of numerous operating utilities in multiple states. At the peak, in 1932, for example, 16 such holding companies controlled fully 75 percent of all power produced in this country (Zardkoohi, 1986, p. 66). These developments raised concerns that such companies might exceed the control of regulators in individual states. Together with the general public's uneasiness about big business during the Depression, this concern led to the passage of the Public Utility Holding Company Act (PUHCA) and the Federal Power Act in 1935.

PUHCA required the massive interstate holding companies that had emerged to register with the Securities and Exchange Commission and to be subject to a variety of structural and operating restrictions. These "registered" holding companies were required to operate in contiguous areas, to engage only in electricity-related activities, and to comply with certain regulatory standards in any transactions among their affiliates. The Federal Power Act was a reflection of the increasing importance of interstate wholesale power transactions. It mandated a uniform system of accounts and gave the newly formed Federal Power Commission (later, the Federal Energy Regulatory Commission) substantial authority over this emerging market.[2]

There were both direct and indirect effects of this legislation and the associated regulation. Directly, they changed industry structure and behavior and thereby undoubtedly influenced aspects of its further evolution. More subtly, they introduced a substantial federal role into the industry that continues to this day.

Yet other trends of the time continued without interruption. Mergers and acquisitions led to the decline in the number of IOUs to 1,400 by 1937, 850 in 1947, and just over 200 at present. By contrast, publicly owned systems grew in number to over 3,000 in the early 1920s before falling to about 2,000 by 1929, a level at which they have since remained. The count of competitive systems declined steadily from about 100 in the 1920s (Emmons, 1993) to the present level of between 15 and 30, depending upon the definition of "competition." The public – private accounting, shown in Table 1.1, does not really capture the relative importance of the two categories of utilities. IOUs have always been fewer in number but of much greater size. By contrast, while there are some large publicly owned utilities — at various times, Los Angeles, Seattle, Detroit, Cleveland, and other cities — their aggregate role in the industry has always been much smaller.

It was during this period of the 1930s that the two other types of utilities made their appearance. The Rural Electrification Administration was established in 1936 to help finance the distribution of electric power to sparsely populated rural areas. Within five years there were more than 800 rural electrical cooperatives, approximately the same figure as at present. Typically small, these coops have never represented more than three percent of generating capacity, although they are responsible for seven – to – eight percent of total power distribution (APPA, 1982, p. 67).

Federal power agencies were also created at this time in order to operate and make use of power from hydroelectric projects built by the Army Corps of Engineers for flood control and navigation. The Tennessee Valley Authority is the largest and perhaps the best known of these projects which provide huge amounts of power essentially as wholesalers to the other sectors of the industry. Federal law requires that publicly owned utilities and rural cooperatives have preferential access (that is, "first claim") to power from these projects, which is much sought after due to its very low cost. It should be noted that their capacity is no longer limited to hydro sources but includes conventional steam and nuclear power units as well. These few power projects represent fully 10 percent of total generating capacity and output in the U.S.

The average size of most types of utilities grew substantially through this period as several advantages of size and integration came to be recognized. Some of these were classical scale economies in generation, that is, larger generating units simply achieved cost savings in construction and in operation (Joskow and Schmalensee, 1986, pp. 48 – 54). Other perceived advantages had their roots in integrated load planning and in coordination of generation, transmission, and distribution. Load planning — conserving on costly capacity to meet varying demand — was facilitated by integrating multiple consuming areas and multiple generation facilities. Coordination among generation, transmission, and distribution implied advantages from vertical integration. The combination of these forces led to an industry with huge integrated utilities as the norm.

1.2 Outline of the Current Industry

At present, electric power in the United States is a $225 billion industry comprised of 3,225 utilities. Each of these utilities performs one or more of the following three functions — power generation, long-distance transmission, and local distribution. Those that do all three are said to be "vertically integrated."

Most electricity in the U.S. is generated in a process that relies upon fossil fuel (coal, natural gas, or oil) combustion to create steam that powers a generator. Considerably less electricity is produced in nuclear power plants, which rely upon nuclear fission to create steam. A smaller amount yet derives from hydroelectric facilities in which water flow directly drives a generator. Small scale gas turbines are employed by most generating utilities to produce power during peak periods.[3] While energy intensive, they are capital conserving and thus ideal for brief periodic use.

Transmission involves the flow of electricity across significant distances, typically from a generating station to a wholesale purchaser such as a local distribution company or industrial user. Minimization of losses from transmission lines requires that this take place at high-voltages, which in turn require substantial capital and secure rights-of-way over long distances.

Local distribution takes this electricity, wherever it is generated, and delivers it at lower voltages to individual customers, including residential, commercial, and many industrial users. As does transmission, the "wires" part of distribution requires substantial capital and rights-of-way down local streets to each customer. Local distribution also encompasses a "sales" or commercial function — the securing of supply, marketing, billing, and related customer services.

Table 1.2 reports that almost 2,000 utilities in the U.S. are publicly owned, while 265 are investor-owned. IOUs represent the very large and highly integrated sector of the industry, accounting for over three-quarters of ultimate customers, sales, and generating capacity. Apart from a handful of federal power projects, the remaining 1,000 utilities are rural cooperatives. The relative frequency of each type of utility and their relative sales are depicted in Figure 1.1.

As noted earlier, the large number of publicly owned utilities obscures the fact that they serve only about 13 percent of all customers and generate an even smaller fraction of total power. The difference between their shares of generation and sales—nine percent versus 13 percent—implies that they must purchase a substantial fraction of their overall power requirements from other utilities — federal power projects where feasible and IOUs when necessary.[4] In fact, a majority of public systems purchase *all* their power and simply distribute it to their customers, typically within a single municipality. This tendency for public systems to be deintegrated (and for deintegrated operations to be publicly owned) is an interesting phenomenon that will be examined later.

The federal role in the electric power industry — and with it, the industry itself — underwent significant change during the 1970s. Concern over energy security and reliability spawned the Public Utilities Regulatory Policies Act of 1978 (PURPA) which required utilities to purchase power from cogenerators and renewable-fuel energy producers at rates determined by FERC. Increasingly, over the past ten years, these so-called "non-utility generators" have come to form the basis for a functioning, if rudimentary, wholesale market in electric power. That is, they represent significant supply sources independent of traditional integrated IOUs.

In some ways more dramatic is the prospect of end-user competition in electric power. FERC actions during the 1980s encouraged the growth of an independent power generating sector. Together with efforts to allow long-distance "wheeling" of power across other utilities' lines, this created an opportunity for large industrial customers to enter into power supply arrangements with sources other than neighboring utilities. The Energy Policy Act of 1992 (EPAct) expanded and accelerated these initiatives by authorizing FERC to require wheeling of power to local distribution systems and other wholesale purchasers. In early 1996 a FERC

Table 1.2. SELECTED ELECTRIC UTILITY DATA, BY OWNERSHIP (1992)					
	PRIVATE	PUBLIC	COOP	FEDERAL	TOTAL
Number	262	2017	943	10	3232
Sales to Ultimate Consumers (mil mwt)	2122	395	206	49	2763
Percent Sales	76.4	14.3	7.5	1.8	100
Revenues from Sales (mil $)	149,016	23,727	14,462	1,276	188,481
Percent Generation	79.2	*	4.6	*	100

Note: * Total of public and federal is 16.3 %.
Sources: 1) DOE, Selected Financial Statistics,
2) Edison Electric Institute, Statistical Yearbook 1992.

order did precisely this, mandating nondiscriminatory open-access tariffs for wholesale purchasers, de facto unbundling of services by vertically integrated utilities that wheel power, and provisions for transmission capacity expansion. These actions signaled the inevitable transformation of upstream stages of the U.S. electric power industry.

Complementary to federal policy are actions in California, Massachusetts, and many other states that would allow not just large users but eventually all final customers to select among alternative power supply sources. Under "retail wheeling," power would continue to be delivered along regulated transmission and distribution lines owned by specified companies, but the actual retail supply function would be decoupled and opened up to competition[5] The "electric power industry" would become fragmented into a disparate collection of companies performing quite different tasks.

1.3 Plan of This Book

The fundamental changes affecting and proposed for the U.S. electric power industry raise important questions about the consequences for cost and price performance. The analysis in this study provides insight into the likely effects of deintegration, retail competition, and public ownership. Important in their own right, these issues also have relevance well beyond the U.S. electric power industry. Ownership changes, deintegration, and competition are central themes of the restructuring of the electricity sectors of many countries and, indeed, of the restructuring of other industries as well. In addition, the great diversity of the U.S. electric power industry allows examination of numerous other issues. These include the effects of power pools, holding companies, incentive regulation, and

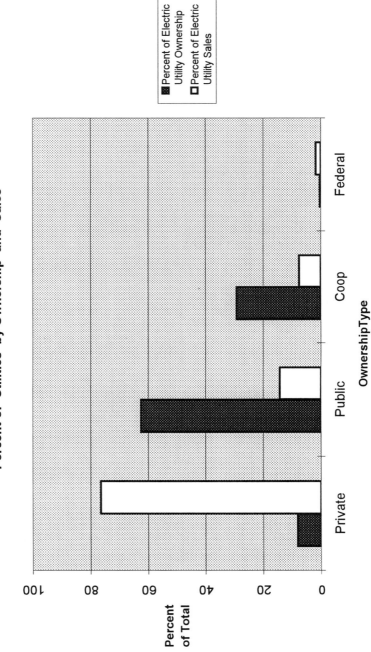

Figure 1.1
Percent of Utilities by Ownership and Sales

combined gas and electric operation on cost and price performance.

The method of analysis in this study is empirical. A substantial data set that accounts for most electric power sold in the U.S. in 1989 is employed. The comprehensive nature of the data, together with the economic modeling, provide answers to such fundamental questions as the effects of public versus private ownership, integration versus deintegration, competition and monopoly, size and scale economies, power pools, holding companies, and other interfirm affiliations, tax and capital cost differences, utility commission characteristics, and differences among customer classes·

We begin in the next two chapters with the principal economic issues. Chapter 2 discusses ownership and regulation, drawing on various models and previous studies as the basis for a preliminary examination of data. Chapter 3 does the same for economies of scale, vertical integration, and distribution competition. Chapter 4 estimates a cost function for U.S. electric utilities and finds that, holding all else constant, publicly owned and investor-owned systems each achieve lower costs in different facets of the electric power business. Public ownership results in superior efficiency in distribution — precisely the circumstance in which it typically arises — whereas IOUs are more efficient at generation, which they dominate. Moreover, economies of vertical integration between the generation and distribution stages are quite important. This fact has significant implications for the overall size of utilities, since vertical economies help to forestall diseconomies of scale that arise in generation and in distribution. It also has implications for the frequency and pattern of public versus private ownership.

Chapter 5 models and estimates the mechanism by which price is set — by the regulator for private firms, and by the municipality for publicly owned entities. Given costs, the price of electricity is found to be lower for three types of utilities — those that are publicly owned, those subject to competition, and those where state utility commissioners are popularly elected as opposed to appointed. Chapter 6 integrates these cost and price functions with a demand function into a simultaneous equations model. This captures the interaction among electric utilities, regulators, and consumers, an interaction that, in fact, jointly determines the market outcome. Reassuringly, this considerably more general approach yields the same substantive conclusions as the simpler model.

A further extension in Chapter 6 disaggregates price effects into residential, commercial, and industrial market segments. Residential customers are identified as the primary beneficiaries of public ownership and of popular election, consistent with the previous observation that publicly owned utilities are end-user (that is, distribution) oriented. These findings, in turn, underlie Chapter 7, which examines the economic and political determinants of mode of ownership and mode of utility commissioner selection. Again, the role of residential customers and certain other predicted characteristics of the market, are found to be important.

In pursuing these questions, the present study takes a significant step beyond existing work. Its data cover far more utilities and in far greater detail than any previous study. The modeling not only replicates existing formulations but also integrates them into a more comprehensive framework. The results represent compelling evidence about the effects of public ownership, the viability of competition, and the reasons for vertical integration in the electric power industry. They go far towards resolving certain inconsistencies in past studies and have considerable relevance to an industry embarking on major transformation.

Notes

1. The term "publicly owned" should not be confused with "publicly held." The latter denotes a firm whose stock is widely owned, as opposed to "closely held" in the hands of a few owners who do not offer it for public sale. Publicly owned enterprises issue no stock. Rather, they are creations of their municipal governments and "owned" by the public through that government.

2. This federal role over interstate, wholesale transactions has remained distinct from state regulation of retail sales. For further discussion, see articles by Joskow, Michaels, Houston, Gordon, and Geddes in *Regulation* (Winter 1992).

3. New combined cycle gas turbines will be discussed later.

4. This dependence upon IOUs for power supply, or alternatively upon adjacent or surrounding IOUs to transmit ("wheel") power from distant sources, has been the cause of constant dispute before regulators and the courts. See, for example, *Otter Tail Power* v. *U.S.* 410 U.S. 366 (1973) and *Town of Concord et al* v. *Boston Edison* 915 F.2nd 17 (1st Circ., 1990), *cert. denied*. In conjunction with the latter, see Kwoka (1992).

5. The original California proposal is contained in the April 1994 California PUC "Order Instituting Rulemaking on Restructuring California's Electric Services Industry and Reforming Regulation." A more modest version, emphasizing a statewide power pool, was ultimately approved in "Order Designating Proposed Policy," September 1995.

6. As will be discussed further, this study covers only IOUs and publicly owned utilities in detail. Power projects and rural coops are fundamentally different. The former are exclusively wholesale generators, with federal ownership and altogether different pricing rules. Rural electric coops are collectively owned by their consumers and by design serve areas that the "market" would avoid or poorly serve.

Chapter 2

PUBLIC VERSUS PRIVATE OWNERSHIP AND REGULATION OF ELECTRIC UTILITIES

The prevailing economic philosophy at the turn of the century in the U.S. emphasized private ownership and laissez-faire competition. Intervention in the sometimes-freewheeling market process, much less outright public ownership, was regarded by most policymakers and economists as anathema. Yet the Progressive movement and a minority of the economics profession at the time argued that "natural monopolies" made competition unworkable and that "such enterprises must be regulated by the State or they must be owned by the State."[1] Of these alternatives, many viewed regulation as costly, ineffective, and frequently corrupt and instead advocated *public* provision of water distribution, natural gas, street railways, and electric power services.

This debate over public ownership, regulation, and competition for utilities has been renewed repeatedly during this century. While much of the early discussion was polemical in nature and reliant on anecdotal evidence,[2] within the past 30 years, there have been greater efforts to formalize the economic issues and to test specific hypotheses about the performance consequences of these market and institutional alternatives. Among these issues, this chapter will focus on the question of public versus private ownership of electric utilities. Economies of scale, vertical integration, and the role of competition will be addressed in the next chapter.

2.1 Economic Models of Public Ownership

Modern economic analysis of public versus private ownership has at different times emphasized property rights, regulatory disincentives, informational asymmetries, and agency problems. Each of these poses the conceptual issue somewhat differently, meriting separate attention.

The first formalization of the underlying economic issues was due to Alchian and Demsetz (1972). In their so-called property rights theory of the firm, the key restraint on managerial discretion and abuse thereof is the ability of informed owners to divest themselves of ownership shares in an enterprise that does not perform efficiently. For corporations with publicly held shares, such divestiture is readily accomplished through the stock market. Stock sell-offs put pressure on current management to improve performance, or else the company becomes an easier target for takeover by owners more dedicated to profit maximization.

Public ownership, by contrast, does not allow for easy divestiture by dissatisfied owners. "Ownership" inheres in residency in a particular municipality and can be "divested" only by physical relocation. The stake of a typical owner — for example, some fraction of his or her electric bill — simply does not warrant undertaking that enormously expensive effort (although the same may not be said for large industrial users). The only alternative is to work through the municipality's bureaucratic and political process, mobilizing a constituency for modifying the behavior of the publicly owned entity. This, too, entails very high cost and uncertain payoff relative to the unilateral act of selling one's stock in a corporation.

The property rights theory of the firm therefore predicts that by severing this restraint, public ownership enables the managers of the enterprise and (perhaps more likely) its political overseers within the municipal government to exercise discretion for their "private" benefit. That is, the public enterprise will be used to grant favors to constituencies that will respond by supporting political incumbency. These favors may arise on either the "income" or "outcome" side. On the income side, the publicly owned entity may pay excessive wages, hire unnecessary workers, or purchase materials and services at above-competitive prices. On the outcome side, it may offer services at below-market rates or provide free ancillary services to favored groups. In either case, the shortfall of revenues relative to incurred costs will be met by transfers of funds or services from the municipality to the publicly owned entity.

Peltzman (1971) and De Alessi (1974) extend this theory and offer some evidence. Observing the average price of power supplied by publicly owned utilities to be *lower* than power from IOUs, they conclude that price concessions must be the chosen method of conferring political benefits and go on to develop corollary hypotheses about public enterprises. For example, they argue that such enterprises can be expected to exhibit lesser price differences among categories of users (since votes are likely to be maximized by equal benefits to customers) and weaker association between costs and prices for particular categories of customers. Some data are offered in support of these hypotheses.

Although he does not develop this alternative hypothesis fully, Peltzman notes that, for purposes of evaluating public ownership, the electric utility industry "does have a possible drawback: private utilities are subject to government price regulation, and regulation can be a substitute for government ownership" (Peltzman, 1971, p. 122). While regulation is designed to control the market power of a natural monopoly, economics has long understood that it induces inefficient behavior by the firm. By reimbursing the firm for its incurred costs, rate-of-return regulation blunts if not eliminates the incentive to conserve on those costs. Accounting systems, oversight, audits, and even occasional disallowances of costs may partially offset these disincentives, but regulation ultimately cannot prevent them. Agencies are at a fundamental and inescapable disadvantage vis-a-vis the regulated firm.[3]

These concerns about regulation were being investigated at the same time as was public ownership. The seminal analysis by Stigler and Friedland (1962) compared the price of electric power in the 1907 – 1937 period in states with and without rate-of-return regulation. Finding little or no difference between the two, they concluded that regulation in practice was ineffective in restraining price below the profit-maximizing level.[4] Thus, it appeared that both regulation of private enterprise *and* public ownership were imperfect.

While their respective deficiencies became well-known, only recently has the *choice* between public ownership and regulation of private enterprise been explicitly modeled. Shapiro and Willig (1990) and K. Schmidt (1993) both set out frameworks in which the regulator is an outsider to the firm, constrained by administrative rules and therefore at an informational disadvantage in its effort to ascertain firm costs and establish prices. Public ownership in principle can alleviate this disadvantage, since owners — including public owners — have rights and access to critical information. In this fashion, public ownership addresses the central informational defect of regulation.

Shapiro and Willig do not contend that this implies the uniform superiority of public ownership, since the rents that accrue to possession of critical information can be used as the public manager desires. In particular, "weak" political systems — those that fail to express society's long-term interests — are poor environments for public ownership, since the rents are likely to be misused. In Schmidt's work, privatization represents a credible commitment to limit public subsidies in the face of excess costs (i.e., it represents a "harder" budget constraint), thereby strengthening the enterprise's cost incentives.[5] Privatization is not a panacea in this model, however, since it is less able to control allocative inefficiency — the output reduction resulting from private market power.

Schmidt's results are framed in terms of incomplete contracts and agency theory. In a further version propounded by Laffont and Tirole (1991), the manager of a publicly owned firm, who is the agent, faces the problem that rational investments are subject to subsequent expropriation by the public owners. For example, the firm may be forced to accept below-market returns, retain excess labor during downturns, etc. Since no contractual commitment can prevent these actions, the enterprise manager underinvests in those assets vulnerable to such misuse.

Laffont and Tirole go on to note that the manager of a regulated firm also has weakened efficiency incentives. That manager is the agent for *two* principals with divergent interests — namely, shareholders and the regulator himself. The resulting conflict, particularly with respect to pricing, undermines efficient operation of the private regulated firm. Their analysis further predicts that the regulator may in fact tolerate some degree of allocative inefficiency in order to improve on the productive efficiency that is also important.[6]

2.2 Empirical Research into the Effects of Ownership

Comparisons of the actual performance of publicly owned versus investor-owned electric utilities have long been conducted. The earliest efforts were essentially crude price comparisons, but it was quickly recognized that many factors other than mode of ownership play a role in pricing. Debate over these factors, the primary among these being costs, was a major focus of studies up until the 1940s. Hausman and Neufeld's survey (1994) notes a hiatus in such work from then until about 1970, when increasingly sophisticated techniques began to be applied.

The first of this new breed of studies were due to Moore (1970), Peltzman (1971), De Alessi (1974), Meyer (1975), and Yunker (1975). Strikingly, all found that publicly owned systems charged significantly *lower* average price, but each offered its own explanation for this result. As previously noted, Peltzman and De Alessi largely dismiss the possibility that this could be due to bona fide cost advantages of public ownership and instead interpret the price advantage as a politically motivated benefit to customers.[7] In a more elaborate model, Moore explicitly estimates cost and demand conditions to infer profit-maximizing prices. His comparison with actual prices leads him to conclude (as did Stigler and Friedland) that price regulation of IOUs is essentially ineffective, but that public systems charge significantly less. Both Meyer and Yunker also find price disparities favoring publicly owned utilities and investigate cost differences as a possible cause. Their results confirm lower costs from public ownership, at varying levels of statistical significance.

The common finding of lower cost from public ownership made it more difficult to dismiss lower price as merely a political benefit. After all, if publicly owned utilities indeed achieve lower cost, then any pricing rule based on costs would result in price differences akin to those observed. On the other hand, the interpretation of this cost difference itself varied enormously. Some attributed the cost advantage of publicly owned utilities to subsidies in the form of tax exemptions, low-cost financing, and differential access to hydro power. Another possibility was that higher IOU costs resulted from the disincentives of rate-of-return regulation and the so-called Averch – Johnson effect in particular.[8] These possible explanations were not carefully examined at the time. Meyer, for example, simply states that with respect to the "higher costs...associated with a regulatory environment, one cannot say whether it is ownership mode per se or a regulatory influence since the implication of the Averch – Johnson result is that regulated firms will not produce at minimum costs" (Meyer, 1975, p. 398).

Most of these early studies used models, data, and methodology that, while improvements over what preceded, left a good deal to be desired. For example, Meyer's examination of costs fails to control for input price differences among utilities. Moreover, all of these studies rely upon fairly small samples, often selected to reduce the influence of potentially confounding factors such as size, self-generation, or type of facility. The consequences for sample properties and statistical significance are not examined.

The next set of studies in the literature involves cost function estimation. The studies reflect developments in both duality theory and in econometric technique, notably, flexible functional forms which permit efficient estimation of production and cost relationships with a minimum of prior restrictions.[9] Most employ the translog functional form and typically focus on the costs of steam power generation only. Studies by Pescatrice and Trapani (1980), Fare, Grosskopf, and Logan (1985), and (in part) Hayashi, Sevier, and Trapani (1987) all report that public systems have lower costs than do IOUs. Pescatrice and Trapani's results suggest a very large and statistically significant effect, ranging from 24 percent to 33 percent. In two other sample years, however, Hayashi et al. find IOU costs lower, while Atkinson and Halvorsen's (1986) results imply equal costs for both types of systems. In a study of distribution only, Neuberg (1977) reports that municipal systems' costs are significantly lower by six percent to 25 percent.[10]

In a further extension of this approach, Hollas and Stansell (1988) estimate a translog profit function for IOUs, municipal utilities, and cooperative systems. They report significant differences among all three types of utilities, with municipal systems the least efficient, followed by coops, and then IOUs. Later work by Hollas, Stansell, and Claggett (1994) compares coops and municipal utilities using a translog cost function, demands by market segment (residential, commercial, and industrial), and pricing for each segment. They do not find any significant effect of municipal ownership on costs; however, residential and commercial prices are lower, while industrial rates are higher under public ownership.

Two recent studies explore subtleties of the cost comparisons. Pollitt

(1995) has examined technical and cost efficiency[11] for samples of utilities and plants in several countries using a number of statistical techniques. For utilities as a whole and also for all generating plants, he finds no significant difference in technical efficiency between those that are privately owned versus publicly owned. For the important base-load plants, however, the latter exhibit significantly technical inefficiency relative to their privately owned counterparts. Furthermore, Pollitt finds no evidence of efficiency differences in the transmission and distribution stages of utility operation that differ by ownership type.

A different variant of the cost phenomenon is examined by Koh, Berg, and Kenny (1996). They find that public ownership is associated with lower costs of electricity generation, but only for *small* systems. For large utilities, private ownership provides lower cost power. The advantage for small publicly owned systems, they speculate, is the result of more effective voter control in small jurisdictions, an issue to which we shall return.

All of these studies are summarized in Table 2.1. As is evident, a preponderance of studies finds lower costs from public ownership rather than from (regulated) private ownership, even after controlling for differences in input prices and many other factors that cause cost differences.[12] More recent studies differ in many respects both from earlier studies and from each other. Among the advances is the effort to identify possible reasons for differential costs typically favoring publicly owned utilities. It is useful to summarize the three leading possibilities, namely, defects of regulation, subsidies to public systems, and end-user orientation.

The first of these — the impact of regulation — argues that public ownership may not so much represent the "ideal," but rather that it is simply a less imperfect governance mechanism than traditional cost-of-service regulation. Some evidence in favor of this interpretation in fact emerges in the cited studies. Where tested, regulated electric utilities are often found to violate the necessary conditions for cost minimization (Atkinson and Halvorsen, 1980, 1984; Hayashi, Sevier, and Trapani, 1985). Moreover, studies comparing performance of public and private provision of such services as waste collection, hospitals, local transit, airlines, and so forth often find different results. In these industries, where strict cost-of-service regulation is not the norm for privately owned firms, the weight of evidence more clearly, though still not universally, favors private ownership (Boardman and Vining,1989; Viscusi, Vernon, and Harrington, 1995). This finding suggests that traditional regulation may be the culprit.

A second possible explanation for the observed cost differential concerns so-called subsidies to publicly owned electric utilities. There are three types of benefits that accrue to public systems: As nonprofit public entities, they are

Table 2.1. Principal Studies of Ownership Effects for U.S. Electric Utilities		
STUDY	**METHODOLOGY**	**FINDING**
Moore (1970)	Demand and cost estimation	Public firms charge 10–22% less than profit-maximizing price, private firms only slightly less (5%).
Peltzman (1971)	Data comparison	Public firms have lower prices.
De Alessi (1974)	Simple price regression	Public firms have lower prices.
Meyer (1975)	Simple cost function	Public firms have significantly lower costs overall and for most cost categories.
Yunker (1975)	Regression cost, price models	Equal costs, but public firms charge less.
Atkinson and Halvoren (1980)	Translog cost function	Equal cost inefficiency (2.4%) from both.
Pescatrice and Trapani (1980)	Translog cost function	Public firms more cost efficient; 24-33% less.
Fare, Grosskopf, and Logan (1984)	Linear programming model	Public firms slightly more efficient.
Hayashi, Sevier, and Trapani (1987)	Translog cost function	Public firms more efficient in 1960s, private firms in 1970s.
Hollas and Stansell (1988)	Translog profit function	Private firms most efficient, public firms least, coops in between.
Hollas, Stansell, and Claggett (1994)	Translog cost function	Equal costs, but public firms' price lower to residential customers.
Pollitt (1995)	Programming and statistical techniques	Equal technical efficiency, but public firms use suboptimal input mix.
Koh, Berg, and Kenny (1996)	Translog cost function	Public firms have lower costs at small scale; private firms at larger scale.

exempt from most state and local taxes (primarily income taxes), although most make "payments in lieu of taxes" to local municipalities. Second, as do local governments, public systems issue debt that is tax-exempt to bondholders and which therefore carries a lower nominal rate of interest. And finally, by law, public systems and rural coops have preferential access to low-cost federal hydro power. While they pay the same price for such power, they rely on it more heavily than do IOUs and thus should face a lower average cost of purchased power.

The argument that these factors more or less fully explain public systems' apparent cost advantage was originally made by Peltzman (1971) and by De Alessi (1974) and more recently by Joskow and Schmalensee (1985) and by Putnam, Hayes, and Bartlett (1985; 1994). The latter in particular attempt to measure directly the value of each of these benefits and conclude that equalizing capital costs, tax rates,and hydro power usage for IOUs, public systems, and rural coops would result in IOU costs declining and becoming the least expensive source of power. This conclusion has been challenged by Laskin (1988), Corazzini (1989), and Ulm (1993), among others, in calculations implying that the actual benefits to public systems explain only part of the sizeable cost differential between ownership types. Moreover, it is argued, that IOUs themselves enjoy some substantial "subsidies" in the form of accelerated depreciation and, until recently, the investment tax credit.

Third and finally, it may be the case that publicly owned utilities have perspectives and motivations which are fundamentally different from those of large integrated IOUs. Public systems may be operated and controlled by citizens of the community which each serves, individuals who view their roles in the context of very local needs. Thus, when small and oriented towards the distribution (end-user) function, publicly owned utilities may achieve performance superior to that provided by investor-owned utilities that serve a larger, more diverse, and farther removed customer population.

The possibility that informational advantages, personal commitment, and/or nonprofit objectives may lead to different behavior by such entities has been noted previously in the literature. Hansmann, for example, discusses a variety of ownership alternatives and concludes that "municipally owned utilities can be seen as a form of consumer cooperative" (Hansmann, 1988, p.297, n.58). Cooperatives also exist in the electric power industry, of course, and are intended to exercise cost and price restraint on behalf of their constituency of local customers. In other contexts as well, nonprofit enterprises have been found to exercise pricing and other restraint (Lynk, 1995).

There is, in short, no real agreement as to the reasons underlying the cost and price differentials that often favor publicly owned electric utilities. But a debate that centers on such possible reasons as efficiency, subsidies, regulatory pricing, and localism is at least amenable to factual analysis. That is the approach underlying the present study.

2.3 Investor-Owned and Publicly Owned Utilities in the U.S.

To supplement the broad outline of the U.S. electric power industry that has already been provided, some details of the operating and performance characteristics of the investor-owned and publicly owned utilities in the present database will be presented. We begin by describing that database.

The data for this study cover 543 electric utilities in the United States.[13] The primary sources of information are Department of Energy Forms 1, 412, and 861, which are submitted annually by major investor-owned and publicly owned electric utilities. Beginning in 1986, DOE data collection for public systems became more comprehensive and required greater consistency with data from IOUs. Further revisions in 1989 added to the quality and utility of the data. It is that year, 1989, for which the present data set is constructed.

Data compilation for this study began in 1990 and was largely completed in 1993, although a few additional items were developed later. Most of the actual collection was performed by staff of the American Public Power Association, at all times under the author's direction and supervision. To ensure accuracy and completeness of all data, countless cross-checks were performed. Raw data forms were examined to fill in necessary information. The Department of Energy, state public utility commissions, and on numerous occasions, individual utilities were contacted to supply, verify, or correct data. In order to obtain further information about particular issues, on three different occasions, surveys were sent out to samples of utilities.

Initially, all 182 major IOUs (those with more than $10 million in annual sales) and all 454 major public systems were included in the database. Several factors ultimately reduced these numbers to 147 IOUs and 396 publicly owned utilities. Since the primary focus of the study is on final service pricing, utilities that exclusively or nearly exclusively generate wholesale power were deleted. While many data omissions and inconsistencies were resolved through the considerable efforts described above, some problems defied resolution and led to deletion of a few (typically small) utilities. The 543 utilities in the final database account for 98 percent of all final sales by IOUs and 83 percent of public systems' sales. The somewhat lesser coverage of public systems reflects their smaller size and the sharply rising marginal cost of securing accurate data on them.

Highlights of composite income statements for the investor-owned and publicly owned utilities in the database are reported in Table 2.2. There is an obvious disparity in the sizes of the two segments, with IOUs generating $145.5 billion in revenue as compared to $19.2 for public systems. Operating expenses represent 80 percent of IOU revenues and 90 percent for public systems, the difference due to the fact that the latter issue no stock and hence make no dividend payments.

About 58 percent of IOU operating revenues and 76 percent of public system revenues go to operation and maintenance. Depreciation and amortization represent a somewhat larger fraction of IOUs' operating revenues — 10.0 percent compared to 7.4 percent — while taxes and related items appear to be far larger for IOUs. IOU tax payments to municipalities, states, and the federal government

Table 2.2. COMPOSITE INCOME STATEMENT FOR UTILITIES IN DATABASE (Selected Items, $ million)						
	PRIVATE			PUBLIC		
	Total	% Revenues	% Expenses	Total	% Revenues	% Expenses
Number of Utilities	147			396		
Operating Revenue	$145,530	100.0		$19,206	100.0	
Operating Expenses:	117,012	80.4	100.0	17,266	90.0	100.0
Operation	73,206	50.3	62.6	13,504	70.3	78.0
Maintenance	10,937	7.5	9.3	1,219	6.4	7.1
Depreciation &Amortization	14,612	10.0	12.5	1,418	7.4	8.2
Taxes, Tax-Like payments	17,052	11.7	14.6	527	2.7	3.0
Net Deferred Taxes, ITC	1205	0.8	1.0	0		
Net Services	0			574	3.0	3.3
MEMO: Operating Income	28,518	19.6		1966	10.2	
Net Interest	14,362	9.9		1843	9.6	
MEMO: Net Income	14,156	9.7		123	0.6	

constitute fully 11.8 percent of their operating revenues. In addition, these utilities make net payments for deferred taxes and for the investment tax credit equal to 0.8 percent of operating revenues, for a total of 12.6 percent in total tax obligations. The relative importance of these components is depicted in Figure 2.1 for public systems and in Figure 2.2 for IOUs.

Regulatory costs for IOUs cannot be fully measured. Reported information covers the costs of utility participation in specific regulatory proceedings, which averages only 0.17 percent of their operating revenues. This item, however, does not reflect whatever portions of the costs of their legal departments or tariffing divisions that may be responsive to regulation. Nor does it capture possible costs of modifying the utility's corporate structure to comply with regulatory requirements, e.g., separate divisions for gas and electric operations. And, of course, it obviously it does not measure Averch – Johnson effects or other cost inefficiencies induced by regulation.

With respect to publicly owned utilities, explicit taxes account for only about 1.4 percent of revenues, but this figure needs to be corrected for certain tax-like payments which they make. Perhaps most obviously, the majority of publicly owned utilities make "payments in lieu of taxes" which are levied by city administrations explicitly as alternatives to normal tax obligations. These payments (sometimes called "tax equivalents") comprise another 1.3 percent of public systems' revenues. In addition, such utilities often make less systematic but substantial "contributions" from their earnings to their municipalities. These contributions or "transfers to the general funds" average another 2.7 percent, for

Figure 2.1
Allocation of Operating Revenue - Public Firms

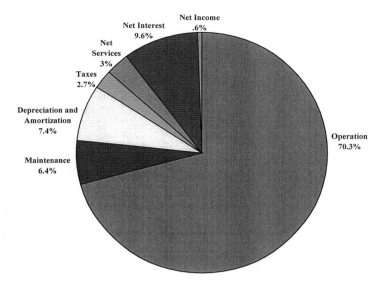

Figure 2.2
Allocation of Operating Revenue - Private Firms

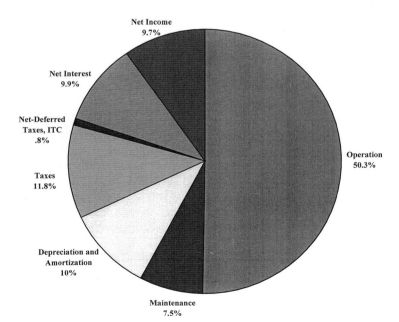

a total of taxes and tax-like payments of 5.4 percent.

It is also worth noting that 81 of the publicly owned utilities in the data set which purchase power from the Tennessee Valley Authority make roundabout tax-like payments which are not included in these totals.[14] Specifically, the price at which they have purchased TVA power includes a payment of five percent that is returned to the states and counties in which the utilities were operating.[15] If added to the other tax and tax-like payments of publicly owned utilities, these would increase the total shown in Table 2.2 by 12 percent and increase the effective percent of taxes and tax-like payments by publicly owned utilities from 5.4 to 6.2 percent.

The relevance of these payments stems from the fact that they are quite explicit alternatives to direct payments by local distribution utilities. That is, those utilities with TVA purchases and payments make correspondingly smaller *direct* tax and tax-like payments to their states and municipalities. This is confirmed by examination of the percentages of revenues accounted for by such payments for those utilities supplied by TVA relative to others which are not. For the 81 public systems purchasing TVA power, their total payments constitute 2.1 percent of operating revenues. The other 315, by contrast, paid 6.2 percent of their revenues.

This large discrepancy implies that accounting only for explicit payments significantly understates the total of taxes and tax-like payments for publicly owned systems. For most subsequent analysis, we nonetheless utilize the Table 2.2 values which do not include TVA payments. There are two reasons for this choice. First, inclusion of TVA payments might raise a question about tax incidence more generally, that is, about the degree to which all public systems and IOUs as well pay prices for inputs that include some measure of tax nominally paid by suppliers but ultimately falling on the utility and its customers. Determining the ultimate burden of all such taxes is beyond the scope of this study, if it is feasible at all. The second reason for not incorporating these payments is a preference for conservative assumptions about publicly owned systems' tax payments in the general analysis. Thus, despite some clear similarities of TVA payments to taxes, the basic approach here is intended to avoid any debatable attribution of taxes paid by public systems.

In addition to these various payments, publicly owned utilities often perform uncompensated or undercompensated services for their municipal governments. Examples include electrical service and use of utility employees, vehicles, equipment, etc. These represent a bit more than 0.3 percent of operating expenses. In turn, municipalities sometimes provide their utilities with free services such as water supply, office space, and accounting services. Despite much speculation to the contrary, the latter prove to be quite modest — less than 0.1 percent of operating expenses. Netting out service flows results in a cost to publicly owned systems of about 0.3 percent (after rounding). The total of taxes and tax-like payments plus net services to their municipalities therefore represents about 5.7 percent of the operating revenues of publicly owned utilities. This should be compared to 12.6 percent for IOUs, still leaving a large differential of 6.9 percentage points.

Further detail on these items is contained in survey of 700 of its members conducted by the American Public Power Association (APPA, 1992). This reported a median value of taxes, tax-like payments, and services as a percent of operating revenues of 5.8 percent, similar to the value found in the present data. As shown in Table 2.3, this percentage increases from 4.8 for the smallest size class to 6.1 percent for the largest utilities. Table 2.4 indicates that almost 75 percent of the total of $943 million in tax-like payments (an amount directly comparable to that in Table 2.2) takes the form of "payments in lieu of taxes." These are paid by 74 percent of 478 utilities completing this part of the survey. The remainder consists of gross receipts taxes[16] plus other taxes and fees, primarily state public utility assessments, property taxes, and franchise fees. Fifty percent of utilities make one or more of these latter types of payments.

Most utilities make payments in lieu of taxes to their local governments, the remainder — particularly utilities in Kentucky and Washington — to state governments. Table 2.5 shows that these payments are most often determined by one of three methods: as a percent of gross operating revenues, as a property tax equivalent, or on an ad hoc or "as needed" basis by the city council or utility board. A different APPA survey found that payment formulas most often apply to larger utilities and to those controlled by utility boards rather than city councils.[17] The alternative criterion of payments "as needed," by contrast, suggests possible use of utility revenues to balance city budgets and/or help fund special projects, reflecting the concern that some have articulated about publicly owned enterprises.

Table 2.4 also shows a total of $61.2 million of free or reduced-cost services reported by utilities as being provided to their municipalities. These consist primarily of electrical services, but also include use of employees, equipment, and supplies. Well over half of all utilities reported providing one or more of these services. This total in turn should be compared to free or subsidized services provided *by* the municipality *to* the utility, as indicated in the last block of Table 2.4. The latter amount — $5.2 million, from 26 percent of surveyed utilities — is obviously much smaller than the value of services *from* utilities, belying the common view that utilities are large net beneficiaries of subsidized services. The largest category of municipally provided services is use of employees, followed by use of vehicles and equipment, free or reduced cost services, and materials and

Table 2.3. TAXES, TAX-LIKE PAYMENTS, AND SERVICES		
Revenue Class (millions)	**No. of Utilities**	**Median Payments (%)**
Less than 2	111	4.8
2 – 5	110	4.5
5 – 10	123	5.2
10 – 20	162	6.0
20 – 50	125	5.9
50 – 100	34	6.1
100 or more	35	6.1
TOTAL	700	5.8

Source: APPA (1994)

Table 2.4. BREAKDOWN OF PAYMENTS AND SERVICES		
	AMOUNT (millions)	**PERCENT**
FROM UTILITY		
TYPE OF PAYMENT		
Payments in lieu of taxes	709.0	75.2
Gross receipt taxes	182.4	19.3
Other taxes and fees	52.1	5.5
TOTAL	943.5	100.0
TYPE OF SERVICE RENDERED		
Electrical services	50.1	81.9
Employees	9.6	15.7
Vehicles, equipment, materials & supplies	1.5	7.4
TOTAL	61.2	100.0
TO UTILITY		
TYPE OF SERVICE RECEIVED		
Employees	3.4	65.7
Vehicles, equipment, materials & supplies	1.3	24.9
Other	0.5	9.4
TOTAL	5.2	100.0

Source: APPA (1994)

Table 2.5. METHOD OF PAYMENT DETERMINATION	
METHOD	PERCENT
Percent of gross revenue	26
Assessment of utility and city budgets	25
Property tax equivalent	21
Percent net utility plant	5
Charge per KWH sold	4
Percent net income	2
Percent operating income	1
Other	16
TOTAL	100

Source: APPA (1995)

supplies.

These data nonetheless confirm that IOUs face a substantially greater payments' burden than do public systems. Yet, it should be emphasized that these utilities differ in many characteristics other than ownership that may contribute to these differences. Table 2.6 documents, for example, that the 147 IOUs in the database average 15,400 mwh annual sales, whereas the 396 publicly owned systems sell less than 1,000 mwh of power each. The t-statistic of 16.40 indicates the high degree of statistical significance to this difference. There are similarly large disparities in their numbers of customers, transmission pole miles, and distribution miles. That said, both segments of the industry are characterized by enormous is substantial overlap between them. The smallest utility in the database is the investor-owned Northwestern Wisconsin Electric, and the largest publicly owned system, the City of Los Angeles, sells 22 million mwh of power, an amount that would place it in the top sales quartile of IOUs.

With respect to other characteristics as well, the two segments exhibit significant differences. While the average utility generates four million mwh of power, 241 generate none at all and one, Commonwealth Edison of Illinois, produces over 80 million mwh. Non-generators are predominantly publicly owned systems — 233, or 58 percent of their total number — plus eight IOUs.[18]

Another factor affecting performance is the type of technology employed in the generating facility. Hydro power has exceedingly low costs, for example, while peaking capacity is by design capital-conserving but relatively fuel-intensive, resulting in high operating costs. Once thought to be low-cost, nuclear costs have

Table 2.6. STRUCTURAL FEATURES OF UTILITIES IN DATABASE				
	TOTAL (Average)	PRIVATE	PUBLIC	t-STATISTICS
NUMBER	543	147	396	-
MWH (1000)	4,837	15,400	918	16.40
CUSTOMERS	172,085	554,643	30,075	14.65
TRANSMISSION POLE MILES	986	2,390	71.3	12.89
DISTRIBUTION MILES	4,542	15,593	732	15.09
GENERATION (1000)	4,020	13,600	459	15.68
MEMO: No Generation	241	8	233	-
VINTAGE	1966.7	1966.6	1966.7	0
CAPACITY TYPE (%)				
Steam	61.0	68.0	55.0	3.21
Nuclear	6.4	10.5	2.9	4.69
Hydro	11.8	9.1	14.1	1.56
Other	20.5	12.2	27.7	4.84
CUSTOMER SEGMENTS (%)				
Residential	35.8	33.3	36.8	3.17
Commercial	25.4	28.3	24.3	2.72
Industrial	35.3	35.3	35.3	0
Resale	3.5	3.1	3.6	0.23

become far more problematic in the past two decades. As Table 2.6 shows, on average 61 percent of capacity is conventional steam generation, six percent is nuclear, 12 percent hydro, and the rest ("other") is largely peaking units.[19] IOUs have significantly more steam and nuclear capacity, reflecting base-load generation, while public systems rely more heavily on hydro and "other" (peaking) capacity.

A further dimension to technology is the vintage of the generating equipment. More recent equipment can be expected to embody the most modern production techniques and to reflect current and expected fuel and labor prices. Table 2.6 reports, however, that public and private systems have essentially identical capital vintages. Measured as the weighted average year in which their generating units were put into service, this is 1966 in both cases.

Further, this table displays the division of final power sales into the three major customer categories — residential, commercial, and industrial uses — plus a small amount devoted to street lighting and the like. Although the percentages for each segment do not differ greatly between publicly owned systems and IOUs, publicly owned utilities do sell significantly more of their power to residential customers and less to commercial users. Industrial and other applications represent essentially equal proportions for both types of power companies. We shall return to some of the implications of these differences in Chapter 7.

Turning next to performance, Table 2.7 reports prices and the principal factor costs for the 147 IOUs and 396 publicly owned systems in the database. Beginning with the former, perhaps the most notable fact concerns average price, which is 5.92 cents per kwh[20] overall. The average price of power sold by privately owned utilities is 6.25 cents, however, as compared to 5.81 for publicly owned systems. The difference of 0.44 cents per kwh or 7.4 percent is statistically significant with a t-value of 2.84, indicating that there is indeed a systematic price difference between publicly owned and privately owned electric utilities.

As noted, the major customer categories are residential, commercial, and industrial. Price to each of these segments is also reported in this table and depicted in Figure 2.3 to determine whether distinctly different patterns emerge and whether the lower price is perhaps limited to just one segment. Clearly, publicly supplied power is far less expensive for residential customers, by 1.56 cents per kwh or 23.4 percent, and substantially cheaper for commercial users, by 0.53 cents or eight percent. Both differences are statistically significant, with t-values of 9.64 and 2.79, respectively. Industrial customers appear to face essentially identical prices regardless of the ownership of their supplying systems. Indeed, any other finding would be suspect, given that industrial users can shop for power from nonlocal suppliers or generate power themselves.

These findings mirror those of most past studies showing lower electric power prices from publicly owned utilities than from IOUs. But, of course, the data in Table 2.7 represent crude, uncontrolled comparisons and therefore simply underscore the need for more sophisticated analysis. A similar observation applies to the cost comparisons reported next. Overall, average costs are almost identical between system types: 5.58 cents per kwh for IOUs and 5.50 cents for publicly owned systems. This equality masks significant differences among major components of overall costs, however. Publicly owned utilities have significantly *higher* costs per kwh of power supply but significantly *lower* overhead costs. Power supply is defined as power generation plus purchases, whereas overhead consists of customer accounts, customer service, sales, and administrative and

Table 2.7. PRICE AND COST COMPARISONS FOR UITILITIES IN DATABASE				
	TOTAL (Average)	PRIVATE	PUBLIC	t-STATISTICS
PRICES (cents/kwh)				
Price-Average	5.92	6.25	5.81	2.84
Price-Residential	6.66	7.80	6.24	9.64
Price-Commercial	6.67	7.06	6.53	2.79
Price-Industrial	5.13	5.13	5.14	.01
COSTS (cents/kwh)				
Cost-Average	5.52	5.58	5.50	.60
Cost-Avg. Supply	3.65	2.98	3.90	5.44
Cost-Avg. Overhead	0.49	0.63	0.44	4.75
INPUT COST				
Price Steam Fuel (cents/kwh steam)	2.21	1.97	2.55	2.94
Price Nuclear Fuel (cents/kwh nuclear)	8.35	7.97	9.55	.74
Price Purchased Power (cents/kwh)	3.90	3.78	3.95	.93
Price Labor ($/hr.)	5.14	5.35	5.06	4.52
Cost Com/Pref/Debt (%)	7.07	8.10	6.69	4.53
Cost Capital (%)	5.73	8.10	5.04	9.26
Tax Rate/Direct (cents/kwh)	0.417	0.712	0.307	9.65
Tax Rate (cents/kwh)	0.433	0.712	0.329	9.21
Tax Rate/TVA (cents/kwh)	0.452	0.713	0.355	8.68

general expenses. These observations suggest a "leaner" structure for publicly owned utilities, lower costs for their distribution function, but a higher cost of actual electricity production.

Some possible explanations for these cost differences emerge in the next block of Table 2.7. Of greatest importance to most utilities are the prices which they must pay for fuel and for purchases of power, items accounting for about 35 percent and 20 percent of operation and maintenance expenses, respectively.Publicly owned utilities appear to pay more for both nuclear and steam

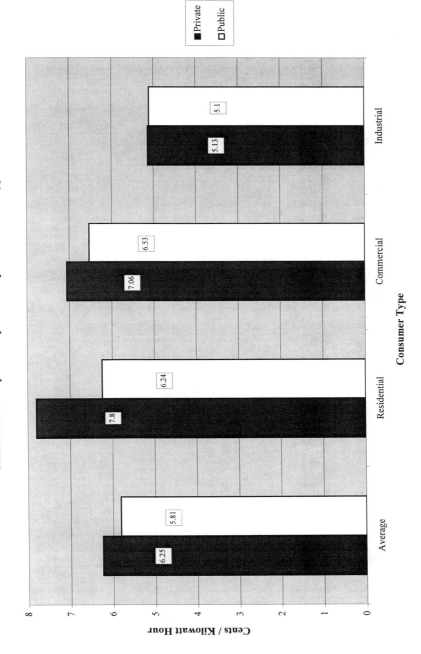

Figure 2.3
Relative Electricity Prices by Ownership and Customer Type

fuel,[21] although only steam fuel expense is significantly greater, with a t-statistic of 2.94. The price which they pay for purchased power also exceeds that paid by IOUs, although this differential also falls short of statistical significance.[22] Lower overhead costs for publicly owned utilities are at least partly explained by the modestly, but significantly, lower hourly wage relative to IOUs.

The fact that public systems purchase power at prices no lower than IOUs runs counter to the view that low-cost federal hydro power is a major advantage to publicly owned utilities (although the present comparison remains crude). There is, however, support in these data for two other supposed advantages to such utilities, namely, capital costs and tax burden. First, with respect to the cost of capital, the conventional measure is the weighted average cost of all types of capital held by a utility: common stock, preferred stock, and long-term debt. The cost of long-term debt is measured as interest on such debt as a percent of long-term debt outstanding. The cost of preferred stock is computed as total preferred dividends divided by the book value of preferred stock, while the cost of common is calculated as the fourth-quarter (1989) dividends paid, annualized and divided by year-end stock price. Since publicly owned systems issue no stock, their cost of capital reduces to their debt cost.

This definition results in the measure of capital cost labeled "Cost Com/Pref/Debt" in Table 2.7. As shown, this averages 7.07 percent for all utilities, but 6.69 percent for publicly owned systems and 8.10 percent for IOUs. The 1.41 percentage point differential represents a statistically significant cost advantage favoring public systems.

A shortcoming of this measure is that the balance sheets of publicly owned systems contain two capital-like items on which they pay no interest. "Investment of municipality" includes such things as the cost of a debt-free utility plant constructed or acquired by the municipality for its utility. "Constructive surplus/deficit" covers similar transfers plus certain supplies provided by departments of municipalities to the utility. A second measure, denoted "Cost Capital," treats these two items as interest-free loans to publicly owned systems, reducing their effective cost of capital to 5.04 percent. The differential with IOUs now increases from 1.41 to 3.06 percentage points. By either measure, publicly owned utilities face substantially lower capital costs relative to IOUs.

With respect to taxes, actual payments for taxes and tax-like obligations are computed differently for IOUs and public systems. For IOUs these are the sum of federal income taxes, other income taxes, other taxes, and a minor adjustment for any negative net credit balance under accelerated depreciation.[23] Public systems' payments are comprised of taxes, taxes equivalents, and "contributions" (that is, transfers to the general municipal fund), as previously discussed. These aggregations are taken as a proportion of total kwh and denoted "Tax Rate/Direct" in Table 2.7. The average of 0.417 cents per kwh combines a rate of 0.712 for IOUs with a rate half that, 0.307, for publicly owned systems. The difference by ownership is statistically highly significant.

This computation should be modified to reflect the previously-noted fact that services performed by publicly owned systems for their municipality are often explicit alternatives to tax payments. Thus, utility services, less services performed by the municipality for the utility, are included in "Tax Rate." Since such services are not overwhelmingly large, this results in a modest increase in the effective tax burden of public systems. Table 2.7 shows the total tax rate for publicly owned utilities now to be 0.329 cents per kwh. This still leaves a large and significant differential between the IOUs' tax rate and that of publicly owned systems.

One possible further modification would account for TVA payments by 81 publicly owned utilities and one IOU. Since their local contributions are adjusted to reflect the contractual pass-through of TVA revenues, those payments might be added to the tax payment total. The result is "Tax Rate/TVA" in Table 2.7, with a rate of 0.713 cents per kwh for IOUs and 0.355 cents for public systems. These represent increases of 0.001 and 0.025 cents per kwh, respectively, over Tax Rate.

Among these alternative measures of both capital costs and tax burden, the least appropriate ones appear to be the very narrow, conventional definitions. In the present context, they fail to capture important and distinctive features of the financial structure of electric power enterprises, both public and private. Consequently, use of them would likely misrepresent actual capital costs and tax burdens. Instead, "Cost Capital" will be employed in the empirical analysis of publicly owned and investor-owned utilities that follows. As between "Tax Rate" and "Tax Rate/TVA," for reasons previously discussed, the former is chosen as the more conservative representation. In the discussion and analysis to follow, these will be referred to as *COSTCAP* and *TAXRATE*.

2.4 Summary

This chapter has reviewed theory and empirical evidence on the issue of public versus private ownership of electric utilities. Theory emphasizes a number of reasons that performance might differ depending upon ownership, but nearly all imply that private ownership should be superior. Previous empirical studies, however, are decidedly more mixed. At least as many studies find that public ownership produces power more cheaply, or prices power less than comparable private systems. Possible reasons for this result include regulation of private systems, subsidies under public ownership, and differences in objectives between the two types.

A preliminary look at the data on the U.S. industry reveal its considerable diversity and complexity. Publicly owned and privately owned electric utilities represent important and viable alternative structures. A number of major differences between them are clear, differences that must and will be taken into account in the analysis to follow.

Notes

1. Richard T. Ely, quoted in Hausman and Neufeld (1991).

2. Or humor: Ambrose Bierce defined a lighthouse as "a tall building on the seashore in which the government maintains a lamp and a friend of a politician" (Bierce, 1957, p. 107).

3. Further discussion of this problem of regulation can be found in such sources as Viscusi, Vernon, and Harrington (1991), Baron (1989), and Pechman (1993). Note that this distortion is distinct from the scenario in which agencies are "captured" by the firms or industries which they supposedly regulate.

4. Similar results were reported by Jarrell (1978). See also Priest (1993).

5. See also Shleifer and Vishny (1994).

6. More recently, Laffont (1995) has sought to model the political process that conditions the results of both regulation and public ownership. Such models will be discussed further in Chapter 7.

7. And where publicly owned systems' costs might have appeared lower, this is rather summarily attributed to subsidies to such systems. See Peltzman (1971) and discussion below.

8. The Averch – Johnson effect predicts that a regulated firm will undertake excess capital investment in order to enlarge its rate base and thereby increase its realized profit. Efforts to demonstrate this effect empirically have met with only mixed success. Nonetheless, this proposition was very influential in the debates over regulation and public ownership. See Averch and Johnson (1962) and, for a review of the evidence, Berg and Tschirhart (1988).

9. For discussion of these advances, see, for example, Berndt (1991).

10. Other cost function studies have examined rural electrical cooperatives, either by themselves (Hollas and Stansell, 1991) or in comparison with IOUs (Berry, 1994). Both conclude that coops are inefficient.

11. Technical efficiency reflects proximity to the production frontier, whereas cost efficiency captures cost minimization given factor prices.

12. See Boardman and Vining (1989) and Peters (1993) for surveys of more studies and different interpretations of their implications.

13. The utilities are listed in Appendix Table A.2. That appendix also contains descriptions and sources of all data, together with a summary table of variable names.

14. There is one IOU — Nantahala (NC) Power & Light — which does so as well. Present discussion focuses on publicly owned systems that account for over 99 percent of the revenues and payments in question.

15. In 1989, these payments totaled $142 million, ranging from $24 million by the City of Memphis to about $300,000 in the case of Etowah, Tennessee. Nine states received payments from TVA, with Tennessee itself accounting for 61 percent of the $232 million total. State governments collected almost all, with counties receiving about one million dollars. See TVA (September 1989).

16. The gross receipts tax, sometimes called a state utility tax, is collected by the utility and remitted to the state or local government.

17. APPA, 1995. Seventy-one percent of utilities with fewer than 500 customers make such payments, as compared to 89 – 96 percent of those with at least 20,000 customers. Seventy-six percent of utility boards employ formulas, as compared to only 47 percent of city-council-governed utilities.

18. These are Eastern Edison (Massachusetts), Kingsport Power (Tennessee), Massachusetts Electric, Mt. Carmel Public Utility (Illinois), Rockland Electric (New Jersey), Southwestern Electric Service (Texas), Union Light Heat and Power (Kentucky), and Wheeling Power (West Virginia).

19. Since these averages are across utilities, they do not represent the percentages of total power produced by steam, nuclear, hydro, and peaking.

20. This may be expressed equivalently as .0592 dollars per kwh, or 59.2 dollars per mwh.

21. As suggested by Table 2.6, steam plus nuclear power represents eighty percent of total capacity. Hydro power has no fuel requirements, and peaking units typically employ gas or similar fuels.

22. These factors, of course, cause the previously noted higher average cost of power supply for public systems.

23. For discussion of this accounting, see the Data Appendix.

Chapter 3

ECONOMIES OF SCALE, VERTICAL INTEGRATION, AND THE ROLE OF COMPETITION

The view that the electric power industry is a natural monopoly is based on its underlying production technology. On their face, generation, transmission, and distribution all appear to be subject to significant economies of scale. The integration of these stages into a single utility reflects a belief in efficiencies from coordination among stages.

Several facets of this standard portrayal, however, are subject to debate. Whatever the scale properties of transmission and distribution, generation economies appear to be exhausted at considerably smaller scale. The latter economies appear to allow multiple generating plants, which might be operated separately from and even compete for downstream load. The efficiency consequences of separating generation from distribution would then depend upon economies of vertical integration. While such economies are widely accepted, there is remarkably little direct evidence as to their existence or magnitude. Even supply competition might be a possibility. Retail competition was once prevalent, and even now there are a number of jurisdictions with duplicative distribution facilities. Since these would seem to violate conventional wisdom about the industry, they bear further examination.

This chapter takes up these various phenomena. We review background studies, set out an analytical framework, and offer some preliminary observations from the present data set.

3.1 Economies of Scale and Vertical Integration in Electric Utilities

Scale economies arise when larger output quantities can be produced with less-than-proportionally greater input quantities. Under these circumstances, a single producer is more cost efficient than an industry comprised of two or more members. Economies of vertical integration are said to occur when successive stages of production can be performed together, that is, by a single producer, more cheaply than by separate producers. Here the trade-off is between the costs of market transactions — arranging for purchase and sale, enforcing contracts, dealing with unanticipated events — versus the limitations of internal, essentially administrative, controls.

The existence of some scale economies in electric power generation, in transmission, and in distribution has never really been disputed. The primary generation technologies of the past half century all involve major pieces of capital equipment whose size and costs have constantly escalated. Conventional steam turbines, nuclear generating plants, and dams for hydro power require large expenditures, long lead times, and complex construction. Recently, however, combined-cycle gas turbine (CCGT) technology has begun to alter the scale properties of power generation.[1]

By contrast, both the transmission and distribution functions represent "wires" businesses. Economies arise in installation (due to construction and rights-of-way requirements), capacity (from the geometric properties of large scale, including minimization of line losses), and operation (network economies). The natural monopoly properties of the wires portion of transmission and distribution are not generally disputed, although the supply function presently bundled with distribution does not have the same economic properties.

Generation costs have been analyzed with increasingly sophisticated econometric techniques during the past 25 years. Studies of plant-level economies are systematically surveyed in Cowing and Smith (1978) and in Joskow and Schmalensee (1983). Both reviews conclude that there are significant scale economies out to a capacity of 400 MW for conventional steam generation and 900 – 1100 MW for nuclear units. These values exceed those characteristic of much of the U.S. electric power industry, but that is in part due to the longevity of smaller scale facilities. Reflecting this, Kamerschen and Thompson (1993) conclude that most steam power increasingly originates in plants of full scale, although nuclear power production continues to face cost savings from scale expansion.[2]

In some contrast to plant-level studies, studies of *firm*-level scale economies have for some time shown that generation economies are indeed exhausted by multiplant companies. The benchmark study in this area is due to Christensen and Greene (1976). Using a sample of 114 electric utilities, they estimate a translog cost function (more on this below) for steam generation and conclude that by 1970, though not in 1955, most firms were at or above minimum efficient firm scale of about 3800 MW. Huettner and Landon's (1978) follow-on study argues that a more pedestrian cost function is better suited than the translog to capturing differences among firms and proceeds to analyze not only generation but also transmission, distribution, and other categories of utility costs in 1971. These researchers ultimately find that scale economies in generation are exhausted at a somewhat lower level than that found by Christensen and Greene. Their results with respect to transmission and distribution are statistically weaker and even sometimes counterintuitive.[3]

Huettner and Landon's study is one of only a few examining the scale properties of functions other than generation. Meyer's earlier study (1975) of public versus private electric utilities did find U-shaped average costs for transmission (although such costs unexpectedly increased from the outset), and unit costs for distribution that initially declined before flattening out.[4] Neuberg (1977) reported evidence of modest economies of scale in distribution over most of the observed size range, with significantly lower costs for publicly owned systems relative to IOUs. Both Henderson (1985) and Roberts (1986) find substantial scale economies in local distribution, Henderson in terms of distribution output and Roberts with respect to customer density but not size of service territory.

In an influential critique, Joskow and Schmalensee (1983) make three points concerning these studies. They express doubts about analyses that focus only on generation, since generation and transmission/distribution are interdependent. That is, power plant size and siting decisions involve technological trade-offs between scale economies from larger generating plants and line losses from longer transmission distances. They also contend that vertical economies from coordinating generation with downstream transmission and distribution make inferences about scale economies in generation per se difficult. Finally, they observe that holding companies, power pools, and a variety of formal and informal agreements affect utilities' operations and costs in ways that such empirical work is unlikely to capture.

Briefly, with respect to this last point, both the Christensen and Greene and the Huettner and Landon studies do address the impact of holding companies. The former study contends that subsidiaries of such companies operate on a fully integrated basis and therefore aggregates their cost to a single observation per *company*. Huettner and Landon by contrast note that not all holding companies are so highly integrated and instead they retain subsidiaries as separate observations. Through the use of dummy variables, this study tests for, rather than assumes, differences between subsidiaries and independent utilities, and reports no evidence of lower costs for holding company subsidiaries.

Power pools, especially so-called "tight" pools which coordinate dispatch, may also serve as an alternative to self-generation of power, conserving in particular on the fixed costs of a utility plant.[5] Christensen and Greene (1978) estimate that simple pool membership results in a cost saving of 4.1 percent and central dispatch an additional 1.3 percent. Neither of these estimates, however, is statistically significant. Rather, it proves to be common ownership of pool members (for example, by a holding company) that is associated with a significant cost savings, which is estimated to be seven percent.

Returning to the issue of vertical integration, Joskow and Schmalensee's cautions would appear to be well placed. Most electric utilities are at least partially integrated, and it may not be sensible to treat their stages as independent operations. After all, economic theory views integration as an explicit alternative to market transactions, to be employed when the latter are too costly (see, e.g., Carlton and Perloff, 1995). Market transactions may be costly for any number of reasons. For example, if production involves specialized assets and sunk costs, buy-sell arrangements are vulnerable to "opportunistic behavior" in which the buyer seeks to extract rents from a seller. This is possible since, once committed to the market, the seller has few alternative outlets for his product or uses for his assets.[6]

A related concern with reliance on the market arises in the presence of small numbers of agents. In the limit, with a monopoly input seller, the buyer will face an inflated price, distorting its input choice and reducing output, all with adverse consequences for economic efficiency. Vertical integration between stages can (but need not) improve matters under these circumstances. In addition, vertical integration may be more efficient when either supply or demand is subject to substantial uncertainty. Production breakdowns, demand fluctuations, and so forth make it difficult to fully specify all contingencies in contracts between separate parties. In the longer term, such uncertainty makes investment riskier than under integrated decision making between stages of production.

A number of observers have noted that the electric power industry has the very properties that give rise to gains from vertical integration (Landon, 1983; Joskow and Schmalensee, 1983). All three stages — generation, transmission, and distribution — are characterized by specialized assets with high sunk costs. Scale economies at the downstream transmission/distribution stage, and to some degree at the generation stage, inevitably imply small numbers (indeed, almost always a single transmission and distribution company) with attendant concerns about monopoly power. Periodic and unpredictable transmission bottlenecks further reduce the number of effective alternative competitors.

Beyond that, the electric power industry and transmission in particular have some unusual properties that affect the feasibility of deintegrated operation (Bohn et al, 1984; DOE, 1989e; Doyle and Maher, 1992). Electricity demand is unpredictable, the commodity is nonstorable, and the transmission medium requires energy balance at all times. This implies, for example, the periodic need for compensatory "reactive power" to maintain electrical stability of the line. Furthermore, power flows over transmission lines in accordance with laws of electrical engineering rather than along some hypothetical "contract path" ostensibly connecting parties to a transaction. The creates what is known as "loop flow," flows of electrical energy onto the lines of non-participants to the business transaction. In economic terms, reactive power is a public good and loop flow an externality.

In the face of such characteristics, vertical integration may be an effective device to achieve least-cost dispatch of generating units, aggregation of diverse load patterns, coordination of shutdowns for maintenance, and system reliability. In the longer run, efficient investment planning by the generation sector requires information about future downstream load, together with a simultaneous decision regarding plant size, siting, and transmission system. For all these reasons, Joskow concludes (1989, p. 139):

> The combination of economies of scale, multiproduct production, and vertical integration provide the primary public interest rationale for the emergence of vertically integrated utilities with defacto legal monopoly franchises to provide retail service to a specific geographical area, subject to price regulation.

What empirical evidence does exist, some of it undertaken in response to the above critique, largely supports this conclusion. Henderson (1985) conducts a formal mathematical test of separability of generation and transmission/distribution costs within a translog framework. By determining that downstream costs are dependent on *input* usage at the generation stage, he concludes that the cost function fails a test for separability.[7] This result is seconded in Roberts (1986).

More direct tests of vertical integration are performed by Kaserman and Mayo (1991) and Gilsdorf (1994, 1995). Each study relies upon a multiproduct cost function to determine economies of vertical integration and economies of scale, and ultimately whether, in conjunction with each other, they imply that electric power is a natural monopoly. The Kaserman – Mayo article offers an innovative adaptation of the conventional multi*product* cost function to test for possible multi*stage* economies between generation and transmission/distribution. The fundamentals of that approach are introduced here, with certain necessary modifications reserved for the next chapter.

It is useful to begin by defining *economies of scope* in a conventional two-product cost function.[8] Denoting outputs Y_1 and Y_2, the costs of producing Y_1 and Y_2 jointly may be expressed $C(Y_1,Y_2)$ and may be compared to the costs of separate production—$C(Y_1,0$ and $C(0,Y_2)$. Then economies of scope are said to exist if

$$C(Y_1, Y_2) < C(Y_1, 0) + C(0, Y_2) \tag{3.1}$$

Such economies may arise for either of two reasons. The cost function may have some indivisible input used in the production of both Y_1 and Y_2, or there may be a cost interaction ("cost complementarity") between the two outputs in the production process. Both of these may be illustrated with trivial cost functions. Letting F represent the cost of an indivisible input,

$$C = F + G + D \tag{3.2}$$

is characterized by economies of scope, since separate production of any positive amount of G and D would have to duplicate F. Cost complementarity is illustrated by

$$C = G + D - G \cdot D. \tag{3.3}$$

Since $G \cdot D$ equals zero for separate production of either G or D, the negative sign implies that joint production is cheaper by the amount of the interaction term.

Economies of vertical integration in electric utilities can be captured by a straightforward extension. Letting G be generation output and D be distribution output, the costs of an integrated utility, $C(G,D)$, are compared to the costs of a pure generator plus those of a pure distributor, i.e., $C(G,0) + C(0,D)$.[9] By analogy to the inequality in (3.1), if the latter sum exceeds the costs of joint production, the production process may be said to exhibit *economies of vertical integration*. The magnitude of such economies may be expressed in percentage terms as follows:

$$S_v = \frac{C(G,0) + C(0,D) - C(G,D)}{C(G,D)} \tag{3.4}$$

A positive sign on S_v indicates such economies, a negative value diseconomies, while its magnitude reveals the extent of either.

This framework provides a simple technique for testing for vertical economies and measuring their magnitude. Kaserman and Mayo use estimates from a quadratic cost function to test the relevant inequality between separate and joint production costs, and to calculate S_v for various combinations of generation and transmission/distribution outputs. They conclude that there are indeed significant efficiencies of integration for most electric utilities, the only exceptions being the smallest size classes for both functions. For the average size utility, the cost savings are found to be approximately 12 percent.

Gilsdorf notes some theoretical advantages of the translog cost function as well as limitations of Kaserman and Mayo's particular formulation of the quadratic. Using cost data adjusted for purchased power (see footnote 8), he estimates a translog cost function on all three stages of production — generation, transmission, and distribution — and tests for cost complementarity. He reports no evidence of cost complementarity. Since economies of scope may also arise from common cost elements, this result does not preclude such economies.

In sum, there appears to be convincing evidence of economies of scale in each individual stage of electricity production, though economies in generation are exhausted considerably earlier than for transmission and distribution. The much sparser evidence regarding economies of vertical integration suggests at least the possibility of cost savings from joint generation and transmission/distribution. Given this latter conclusion, it is useful to recall from Table 2.6 that a majority of the 543 utilities in the present database are in fact integrated between generation and distribution. Yet, fully 241 utilities (about 45 percent) are pure distribution utilities, simply purchasing and reselling power. Interestingly, all but eight of the latter are publicly owned. This pattern of industry structure will ultimately need to be reconciled with the apparent advantages of integration.

3.2 Competition in Distribution

Substantial scale economies in the electric power industry have traditionally left little role for conventional competition. Generators have not been able to seek out potential buyers and, reciprocally, customers have had very limited opportunity to obtain supply from sources other than their local franchised utility. While constraints on such transactions are rapidly eroding, here we are concerned with the competition that has already existed.

Past competition has consisted of franchise competition, benchmark competition, wholesale competition, and competition for large industrial customers (Joskow and Schmalensee, 1985, Ch. 2; Owen and Frankena, 1994, Ch. 7). Franchise competition refers to the periodic process in which one utility is chosen from a group of "bidders" for the right to supply a particular service territory. Benchmark or "yardstick" competition involves reliance on the performance characteristics of other utilities as a benchmark against which to compare the one in question. Few contend that either of these constitutes an effective substitute for actual competition, although benchmark competition undoubtedly helps to inform regulators in their efforts to control utility behavior.

Both competition for large customers and wholesale competition are of some importance in the electric power industry, at least in the long run. Competition for large customers focuses on industrial users who may be mobile at least periodically and thus able to seek the most advantageous prices. Wholesale competition has long existed in the form of utilities purchasing power from each other in order to balance their loads, but the amounts involved have been modest and the resulting "competition" muted. More significant, perhaps, is the post-PURPA phenomenon of nonutility generators (NUGs), particularly independent power producers. The latter are pure generators with plants constructed for and output contracted by other utilities with larger distribution requirements than can be supplied from their own generating plants. At present, NUGs represent more than eight percent of total U.S. generation capacity and 40 percent of additions to such capacity in recent years (DOE, 1994).

Despite its importance in the history of the industry and the continued existence of a number of such jurisdictions, *direct* competition between distribution utilities has not attracted very much attention. Hellman (1972) conducted an early examination of these seemingly anomalous cases, offering a number of case studies and concluding that competition from municipal entities lowered prices charged by IOUs. The most extensive analyses, however, are due to Primeaux, who compared prices and costs in subsets of about 50 jurisdictions that appeared to have duopoly utilities. Using a matched sample of 23 such duopolies with other (monopoly) utilities and holding several other factors constant, Primeaux (1977) reported that operating costs for all competitive systems were on average 11 percent less. An alternative specification suggests a varying differential, smaller for smaller utilities but growing with utility size. In a subsequent study of prices, Primeaux (1985) found the average price of municipal utilities in 20 duopoly jurisdictions to be fully 33 percent below comparable utility levels.[10]

Primeaux interprets these results as evidence of cost-inefficiency by noncompetitive utilities, especially the larger ones. He concludes that large monopoly systems may have lower costs in principle, but that these are realized only through the forces of competition. This observation calls into question the actual benefits from scale economies and monopoly franchises. Although Primeaux's data, methodology, and interpretation have all been subject to criticism (e.g., Joskow and Schmalensee, 1985, pp. 61 − 63), the fact remains that these duopoly cases are anomalies worth examining. Whatever their limitations, certainly Primeaux's results provide no grounds for dismissing possible differences in performance outcomes.[11]

In the present study, an extensive effort was made to identify competitive distribution utilities in general and ultimately among those in the database. Beginning with Hellman's and Primeaux's somewhat dated lists of duopoly systems and supplemented with information from *Typical Electric Bills* (DOE, 1988) and other public sources, an initial compilation of candidate utilities was constructed. Most of these clearly involve cases where two utilities, generally an IOU and a publicly owned system, serve parts of a city or town but are constrained to fixed and exclusive territories. At a first approximation, one might not expect much "competition" in such cases, which resemble the fixed boundaries between any utilities. On the other hand, if the boundary cuts directly through a city or town that owns its own utility, that jurisdiction might well pay closer attention to relative rates and "competition" might therefore impose some restraint.

Since publicly available data were often inadequate to classify the remaining utilities, and also to develop additional information for cases of bona fide competition, a questionnaire was sent in 1995 to a total of 80 IOUs and publicly owned systems in 40 jurisdictions where competition was known or thought possibly to exist. Initial responses plus follow-up contacts resulted in 46 returned questionnaires. At least one response was received from almost all jurisdictions, and from all where there was good a priori reason to believe that competition probably existed. Based on this information, a number of additional cases were identified and a final categorization prepared.

All jurisdictions with any nonstandard utility structures are listed in Tables 3.1 and 3.2. Table 3.1 includes those where two electric utilities serve the city or town but with fixed and exclusive service territories. There were 46 such cases in 1989.[12] Table 3.2 enumerates a select group of 22 *other* cities and towns in which at least some degree of actual retail competition existed. That is, in these cases, at least some consumers under certain circumstances were able to choose or switch suppliers.[13]

This table provides further distinctions about the nature of "competition" in these 22 jurisdictions. Twelve of these 22 cities allow *existing* customers to switch electric power suppliers, the most thoroughgoing version of utility competition. *New* residential and industrial users are permitted an initial choice between suppliers in those 12 plus five additional cities, but no subsequent changes in suppliers are allowed. One might expect more muted competitive effects in this last case.

The survey developed additional information about utilities and their customers in the most competitive 12 cities and towns. As summarized in Table 3.3, customers intending to switch are obliged in all cases to notify the new utility, as well as in seven cases to notify their previous supplier. Switches are consummated in a time frame ranging from one to 30 days, with half of the cases requiring at most seven days. Apart from a deposit, no utility imposed a charge for switching and only two had minimum stay requirements.

These facts and conditions would seem to indicate a substantial potential for customers in these cities and towns to switch between suppliers. Yet, the median percent of residential customers who switch in any year is apparently only about 0.3 percent.[14] This percentage ranges from near zero to a maximum of about six percent in Duncan (Ohio) and Floydada (Texas). Several utilities noted in their

Table 3.1. TOWNS WITH MULTIPLE ELECTRICAL UTILITIES WITH EXCLUSIVE TERRITITORIES	
Ajo, AZ	Mishawaka, IN
Catalina, AZ	Holland, MI
Chandler, AZ	Paw Paw, MI
Gilbert Town, AZ	Zeeland, MI
Glendale, AZ	Champlin, MN
Mesa, AZ	Coon Rapids, MN
Paradise Valley, AZ	Vineland, NJ
Peroria, AZ	Endicott, NY
Phoenix, AZ	Jamestown West, NY
Scottsdale, AZ	High Point, NC
Tempe, AZ	Waynesville, NC
Wickenburg, AZ	Amherst, OH
North Little Rock, AR	Sioux Falls, SD
Sherwood, AR	Electra, TX
Cotton, CA	Garland, TX
Security, CO	Liberty, TX
Norwalk, CT	Lakes District, WA
Gifford, FL	Opportunity, WA
Palm Springs, FL	Parkland, WA
Starke, FL	Spanaway, WA
Cahokia, IL	New Martinsville, WV
Centreville, IL	Combined Locks, WI
East Alton, IL	Harford, WI

responses that price competition was blunted by regulatory or municipal boards that governed prices closely despite the presence of a competitive system. Competition with respect to *service* (that is, nonprice) aspects may not be as effective a device for attracting customers.

Competitive utilities were also asked about other possible dimensions of

Table 3.2. CITIES WITH MULTIPLE ELECTRIC UTILITIES AND SOME CUSTOMER SWITCHING						
City/Town	State	Current Customers	New Residents	New Industrial	Exclusive Territory	Notes
Alexander City	AL			X		F
Bay City	MI		X	X		A, B
Bushnell	IL	X	X	X		
Cleveland	OH	X	X	X		
Columbus	OH	X	X	X		
Culpeper	VA	X	X	X		
Dowagiac	MI	X	X	X		
Duncan	OH	X	X	X		
Floydada	TX	X	X	X		
Greer	SC		X	X		
High Point	NC				X	D
Houma	LA		X			
Kennett	MO				X	E
Kirkwood	MO				X	E
Lubbock	TX	X	X	X		
Paris	KY	X	X	X		
Newton Falls	OH	X	X	X		
Piqua	OH	X	X	X		
Poplar Bluff	MO		X	X		
Sikeston	MO				X	E
Traverse City	MI	X	X	X		C
Trenton	MO				X	E

A = Only areas contiguous to Bay City.
B = City firm purchased IOU's equipment inside city limits in 1992.
C = 1994 Agreement between city and IOU effectively ceased competition.
D = Only those customers on the edge of an exclusive service area may choose between firms.
E = 1990 Missouri Legislation effectively ceased competition.
F = Minimum power usage required to switch.

Table 3.3. CITIES WITH MULTIPLE ELECTRIC UTILITIES WHERE CURRENT CUSTOMERS CAN SWITCH							Shared			One Firm Purchase Power From Other	Notes	Public Firm Market Share
City/Town	State	Advance Notice to Switch	Time Length to Change	Charge	Minimum Stay	% Switch/ Year	Poles	Wires	Drops			
Bushnell	IL	Both Firms	2 Days	None	12 Months	0.84%	N	N	N	Y	D	68%
Cleveland	OH	Both Firms	2 Weeks	None	None	0.66%	N	Y	N	Y	C	20%
Columbus	OH	Both Firms	30 Days	None	12 Months	0.03%	Y	Y	N	Y	B	2%
Culpeper	VA	Both Firms	7 - 30 Days	None	None	0.08%	Y	N	N	Y	B	59%
Dowagiac	MI	Both Firms	7 - 30 Days	$50 Deposit	None	0.10%	Y	N	N	Y	B	85%
Duncan	OK	Both Firms	1 Day	Deposit	None	5.92%	N	N	N	N	C	69%
Floydada	TX	New Firms	3 Days	None	None	6.19%	N	N	N	Y	D	60%
Lubbock	TX	New Firms	3 Days	None	None	2.40%	N	N	N	Y	D	54%
Newton Falls	OH	New Firms	7 - 30 Days	None	None	1.30%	N	N	N	N	B	97%
Paris	KY	New Firms	1 Day	None	None	0.27%	Y	N	N	Y	B	57%
Piqua	OH	Both Firms	1 - 2 Weeks	None	None	0.11%	N	N	N	Y	B	98%
Traverse City	MI	New Firms	3 - 7 Weeks	None	None	0.22%	Y	N	N	Y	A,D	90%

Notes: A = 1994 agreement effectively ceased competition.

B = Only public firm responded to survey. Switch data is only an estimate.

C = Only IOU responded to survey. Switch data is only an estimate.

D = Both IOU and public firm responded to survey.

their association. Five of the 12 indicated that they share poles with each other, most on some contractual basis, although one relies upon a "gentlemen's agreement." Two of these five reported that they also share wires, while none do so with respect to drop lines to individual residences. In all but two cases, the utilities are involved in some arrangement to buy and sell power between themselves, and in one case they jointly provide emergency service to customers affected by outages.

Rather than sharing facilities, several utilities instead duplicate them in the competitive area. Perhaps the best known case is Cleveland, Ohio, where Cleveland Public Power (CPP) and Cleveland Electric Illuminating (CEI) have competed since 1906 (Reinemer, 1987). CEI covers the entire city, while CPP currently has duplicate poles and lines in about half of it but plans to expand into the remainder. In the competitive area, CPP attains a 40 – 50 percent share and further claims an annual migration of as much as 1,000 customers from CEI. CPP's rates are said to be 22 percent lower than those of CEI. The latter has responded by selectively reducing its own rates to Cleveland customers by 10 – 15 percent below non-city rates and redoubled its service and marketing efforts.

Other competitive jurisdictions illustrate a wide variety of patterns to their facilities use, from close cooperation and sharing to complete duplication.[15] Brief descriptions of each of these cases, based on information from the survey as well as from other sources, are provided in Appendix B. As all of this makes clear, competition is in fact quite rare, even more so than past compilations have suggested. Bona fide competition in the sense of actual customer switching occurs in only a dozen out of perhaps 2,500 jurisdictions in the country. A few additional cities allow competition for new hookups while several others have "border competition." The ability of these experiences to cast light on the potential of retail competition in the industry nonetheless warrants attention in the analysis to follow.

3.3 Utility Structure and Costs: Some Preliminary Insights

A preliminary examination of the data set used throughout this study provides insights into the effects of vertical integration and competition on the operating costs of electric utilities. Major operating cost items are calculated per unit of final output, that is, kwh of distribution.[16] Unit costs for generation, power purchases, transmission, distribution, and overhead are reported in Table 3.4. The first column notes the averages for various cost items for all 543 utilities. Block A compares integrated versus nonintegrated (i.e., pure distribution) utilities, while Block B compares monopoly versus duopoly distribution systems, and for comparison purposes Block C reviews privately owned versus publicly owned utilities. Within

Table 3.4. MAJOR COST ITEMS FOR UTILITIES IN DATABASE (Cents/kwh)

	Overall Average	(A)			(B)			(C)		
		Some Generation	Pure Distribution	t-Statistic	Monopoly	Duopoly	t-Statistic	IOU	Public	t-Statistic
Number	543	302	241	-	529	14	-	147	396	-
Operating Expenses	5.43	5.27	5.64	2.79	5.43	5.39	0.10	5.29	5.48	1.27
O&M	4.63	4.22	5.14	7.38	4.64	4.26	0.94	4.00	4.86	6.15
Supply Expenses	3.76	3.30	4.35	8.90	3.78	3.12	1.68	3.01	4.04	7.67
Generation	0.81	1.44	-	-	0.78	1.69	3.27	1.73	0.46	15.0
Purchases	2.92	1.82	4.29	17.6	2.96	1.31	3.03	1.29	3.52	13.0
T&D Expenses	0.375	0.349	0.408	1.22	0.373	0.453	0.53	0.354	0.383	0.54
Transmission	0.049	0.074	0.018	7.79	0.049	0.045	0.18	0.094	0.033	7.60
Distribution	3.26	0.275	0.39	2.41	0.324	0.408	0.56	0.26	0.35	1.69
Overhead	0.491	0.578	0.383	5.35	0.486	0.687	1.72	0.633	0.439	4.75
Customer Acc.	0.098	0.109	0.085	3.92	0.098	0.099	0.05	0.14	0.083	8.4
Customer Service	0.025	0.034	0.013	3.03	0.025	0.028	0.15	0.04	0.019	2.62
Sales Expenses	0.007	0.008	0.005	1.45	0.007	0.014	1.06	0.009	0.006	1.38
A&G	0.362	0.426	0.28	4.40	0.356	0.546	1.79	0.445	0.331	3.05

each block, t-statistics are reported to assess the statistical significance of any differences found in the values.

The mean operating expense per kwh for all utilities is 5.43 cents per kwh. As shown in Figure 3.1, the mean for pure distribution utilities is somewhat larger. Although this comparison holds nothing else constant (notably, system size), it is consistent with a cost advantage from vertical integration. Previous analysis summarized in Table 2.2 has shown that the lion's share of Operating Expenses consists of Operation and Maintenance (O & M) expenses. The residual — denoted "Other" in Figure 3.1 — consists of depreciation, taxes, and the like.

O & M expenses are responsible for all (in fact, more than all) of the penalty in Operating Costs faced by pure distribution systems relative to those that engage in some distribution. Differences in O & M expenses are, in turn, a reflection of their largest component. Power supply costs[17] constitute 75 percent of total O & M and average 4.35 cents for distribution systems, as compared to 3.30 cents for integrated utilities. By definition, of course, the latter have substantial generation costs that distribution utilities do not, and so their generation versus purchase costs do not really illuminate the question of relative efficiency. Nonetheless, generation plus purchase by integrated utilities results in significantly lower supply costs than does pure purchase by distribution systems.

Transmission and distribution expenses are much smaller overall (eight percent of O & M) and differ only modestly between system types: 0.275 cents for integrated utilities versus 0.390 for distribution systems. Transmission (the smaller of the two) is more costly for integrated systems, while distribution is costlier for pure-distribution utilities. Overhead expenses are significantly greater for integrated systems than for distribution utilities — about 0.58 cents per kwh versus 0.38 cents. This difference is the cumulative effect of higher costs for customer accounts, customer service, sales, and (most importantly, since it accounts for 74 percent of overhead) administrative and general expenses (A & G) for integrated systems. Most of these differences are statistically significant, but their modest size results in only a small offset to the higher costs of supply, transmission, and distribution faced by pure distribution utilities.

In sum, integrated utilities have lower overall unit costs, with substantially smaller supply and T & D expenses offsetting modestly larger overhead expenses. Block B replicates these comparisons for monopoly versus duopoly utilities and Figure 3.2 portrays the results. Overall operating expenses are slightly but insignificantly smaller for the 14 duopoly systems in the database. While their power supply costs are lower, distribution and overhead costs are larger, undoubtedly reflecting diseconomies associated with duplication of distribution plant and with lower customer density.

Most of the individual cost items are similar in magnitude for monopoly

Figure 3.1

**Operating Costs and Components for
Integrated and Unintegrated Utilities**

(Cents/kwh)

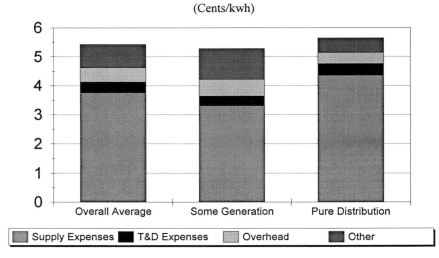

and duopoly utilities. Two interesting exceptions are the costs of generation and of purchase, where there are significant differences. These undoubtedly reflect the fact that duopoly utilities include many pure-distribution systems that have already been shown to have fundamentally different mixes of costs by power source. For this reason, attribution of separate significance to monopoly versus duopoly is almost surely unwarranted.

The final block of Table 3.4 returns to the issue of public versus private ownership. Key comparisons are depicted in Figure 3.3. Operating costs in cents per kwh sold average 5.29 for IOUs and 5.48 for publicly owned systems, although this difference is not statistically significant. On the other hand, the difference in O & M expenses — four cents per kwh for IOUs compared to 4.86 cents for publicly owned utilities — is highly significant, with a t-value of 6.15. As in previous comparisons, the costs of power supply largely account for this difference, with a one-cent differential between the two. This large and significant differential between public and private systems' supply expenses is important to understanding overall cost differences by system ownership.

Within power supply costs, differences are now predictable. Given the preponderance of distribution-only utilities among publicly owned systems, their

Figure 3.2

Operating Costs and Components for
Monopoly and Duopoly Utilities

(Cents/kwh)

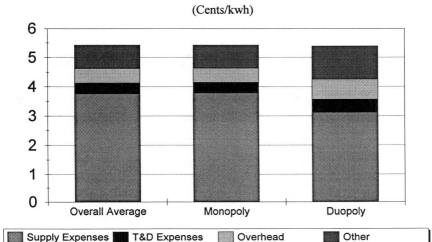

generation costs per kwh sold are considerably lower, while their purchased power costs are correspondingly higher. Nothing can be inferred about the efficiency of their operations from this arithmetic comparison, however.

Total T & D does not differ significantly between IOUs and public systems. This similar total, however, masks differences in the costs of transmission and distribution. IOUs' transmission expenses are nearly 0.1 cent per kwh, while public systems are only one-third as large. This significant difference undoubtedly reflects the much greater degree to which IOUs generate and transmit power for themselves and others.[18] Distribution costs are larger in total, larger for publicly owned systems, but not really significantly so.

Finally, overhead averages about 0.63 cents for IOUs versus 0.44 cents for public systems. Much of the difference is due to the largest component of overhead, namely, administrative and general expenses. A & G averages 0.44 cents for IOUs, as compared to 0.33 for public systems, and indeed, privately owned utilities incur higher costs for each of the other components of overhead for an additional differential of 0.08 cents. In sum, publicly owned utilities appear to

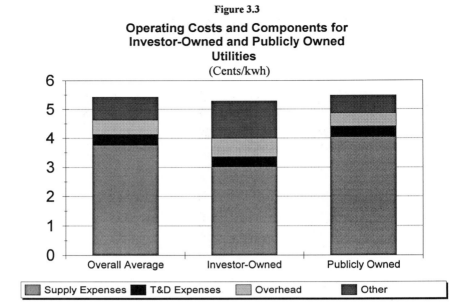

Figure 3.3

**Operating Costs and Components for
Investor-Owned and Publicly Owned
Utilities**
(Cents/kwh)

do a better job of conserving on overhead costs than do IOUs, but with respect to power supply, their costs are higher.

3.4 Summary

Although it is commonplace to assert that the electric power industry is subject to scale economies, this claim is too simplistic. The industry in fact consists of three stages — generation, transmission, and distribution — each with its own scale properties, and with possible vertical economies among them. These various dimensions of costs require must be distinguished and separately examined.

There is an extensive literature on scale economies, but only a few efforts to measure vertical economies in electric power. These studies make clear the challenge of modeling of the multiple dimensions of scale in the industry. A preliminary, simple comparison of costs between integrated and unintegrated utilities, between monopoly and duopoly distribution systems, and between public versus private ownership emphasizes the important differences in cost structures.

Notes

1. CCGT recaptures the heat energy of steam that has been used to drive a turbine and re-uses it in the production process. This results in dramatically higher thermal efficiency relative to conventional steam generation. Together with low levels of pollutants and much smaller scale requirements, CCGT represents an important innovation and has been quickly adopted by independent power producers.

2. Another study, by Krautmann and Solow (1988), suggests that some of the differences in estimates of nuclear plant scale may be due to the very different scale properties of single versus dual reactor plants.

3. They report, for example, some inverted-U-shaped average cost curves.

4. His cost function was fairly ad hoc, however, and failed to control for factor costs and other attributes of utility operations. It might also be noted that Meyer found that the public-private ownership distinction mattered much less for transmission and distribution than for generation.

5. Cramer and Tschirhart (1983) point out that pool agreements and operation have costs as well, leaving net benefits to utilities an open question. Gegax and Tschirhart (1984) model pools as a form of inter-firm cooperation and seek to determine what causes utilities to join.

6. Of course, expectation of this ex post strategy makes the seller less likely to undertake the necessary investment and production ex ante, creating a different market failure.

7. Mathematical separability of inputs and outputs in the cost function is not quite the same as economies of vertical integration. As will be discussed, it is output-output interactions that represent the essential characteristic.

8. The most lucid description of the various cost concepts discussed in this chapter continues to be Bailey and Friedlaender (1982).

9. The accounting costs of a pure distributor will include the costs of purchased power, which is the alternative to self-generation. As will be discussed later, this requires an adjustment to nominal cost data.

10. Only municipal utilities were employed, since the duopoly IOUs typically served much wider territories. This made it difficult to isolate the price effects in a particular city or town.

11. One additional study suggests that among publicly owned utilities, competitive firms have higher variable costs than monopolies, but even its author has some difficulty interpreting this result. (Nelson, 1990)

12. This list focuses on jurisdictions in which an IOU and a municipal utility share the service territory. Cases where the boundary between two IOUs happens to cross through a city or town is less likely to result in measurable market effects than an IOU-municipal boundary. Consequently, two-IOU cases are not considered further here, but can be identified in *Typical Electric Bills*.

13. Table 3.2 does contain five cities in which effective competition existed as of 1989 but ceased thereafter — namely, Bay City and Traverse City (Michigan), plus Kirkwood, Sikeston, and Trenton (Missouri). In Missouri, 1990 legislation required reimbursement to the losing utility for any customers acquired by its competitor.

14. These figures are drawn from survey responses, but the quest ion posed appears to have been answered in somewhat different ways, e.g., as a percent of own customers versus all customers in the town (which may or may not be identical with the competitive service territory). These data are based on the best reconciliation of responses from one or both parties together with outside information.

15. It should be noted that duplicate facilities also may occur when jurisdictions switch from being supplied by one type of utility to the other, but are unable to negotiate a price for sale of existing facilities. Experiences such as that in Clyde, Ohio, will be discussed in Chapter 7.

16. For the present, the focus is on costs of operations rather than on such items as interest

on long-term debt, depreciation and amortization, dividend payments, and net income.

17. DOE terms this "production", even though it includes purchase as one of its components. To minimize confusion, we avoid the term "production" and label purchase plus generation as "power supply".

18. Publicly owned systems' payments for transmission are implicit in their purchased power expenses. This is one illustration of the bases for Joskow and Schmalensee's caution against examination of generation costs in isolation.

Chapter 4

COSTS AND THEIR DETERMINANTS

An appropriate starting point for a comprehensive analysis of the electric utility industry is with costs and their determinants. There are two reasons for beginning with this topic. First, costs are an inherently important dimension of performance, and one for which the effects of ownership, integration, and competition are of considerable interest. In addition, costs are a major determinant of the other facet of performance, namely, prices. An understanding of costs therefore provides the foundation for the price analysis that follows.

This chapter investigates cost determination in a single-equation framework, consistent with most past studies. This approach permits close attention to various cost issues, leaving for a later chapter the development of a simultaneous equations model of the cost-price nexus.

4.1 The Choice of a Cost Function

Utility costs are obviously affected by a myriad of factors. Central to the present inquiry are type of ownership, scale of operation, vertical integration, and competition. Fuel and other input costs, type of generating plant, holding company membership, affiliation with pools, and customer size are among many other influences suggested by engineering, economics, and past analyses. Systematic evaluation of these factors requires econometric analysis, guided by the economic theory of production and costs. We begin by setting out the underlying economic theory.

In general, any cost function takes the form

$$C = C(Q, W, Z) + e, \tag{4.1}$$

where C denotes the cost of producing output Q, W represents a vector of input prices, Z is a set of cost shifters, and e the random error term. To be a bona fide and practical cost function, C should possess certain properties. Baumol et al. (1982) propose four criteria. First, the function should behave in accordance with the economic theory of costs, that is, it should be nonnegative, nondecreasing, concave, and linearly homogeneous in input prices. Second, if the production process involves multiple products, the cost function should be capable of estimation for zero values of some outputs. Third, the function should be consistent with both the presence and absence of significant cost properties. That is, it should allow for economies or diseconomies of scale, economies or diseconomies of scope, cost subadditivity or superadditivity, etc., rather than imposing conditions that preclude the emergence of either alternative. Lastly, the cost function should be parsimonious in the number of parameters to be estimated.

Not surprisingly, no single functional form satisfies all of these conditions best. Two functional forms commonly used in the literature on cost function estimation are the translog and the quadratic. Each has somewhat different strengths and limitations.[1] The translog is a second-order approximation to any arbitrary cost function, minimizing a priori restrictions and allowing elasticities among inputs to vary. For present purposes, however, the limitations of the translog are more problematic. To see this, we set out an explicit form of the translog, for convenience focusing on output Q plus a vector of shifters Z_i:

$$\ln C = \alpha_0 + \alpha_1 \ln Q + \sum_i^n \beta_i \ln Z_i + (\tfrac{1}{2})\, \delta\, (\ln Q)^2 \qquad (4.2)$$
$$+ (\tfrac{1}{2}) \sum_i^n \sum_j^n \gamma_{ij} \ln Z_i \ln Z_j$$
$$+ \sum_j^n \rho_j \ln Q\, \ln Z_j.$$

The key feature of the translog is that it requires taking logarithms of all variables. Since the log of a zero-value variable is undefined, the translog form does not easily deal with continuous variables that may take on zero values or with fixed-effects variables that necessarily equal zero for some observations. In the present case variables with these properties include output Q itself, some control variables Z such as ownership and competition, plus other factors to be enumerated below[2]

These features of the data make estimation and use of the translog cost function difficult in the present context, and instead suggest the quadratic. Again focusing on output Q and a vector of shifters Z_i, a truncated version of the quadratic may be written as follows:

$$C = \alpha_0 + \alpha_1 Q + (\tfrac{1}{2})\, \delta\, Q^2 + \sum_i^n \beta_i Z_i \qquad (4.3)$$
$$+ (\tfrac{1}{2}) \sum_i^n \sum_{j,n} \gamma_{ij} Z_i Z_j + \sum_j^n \rho_j Q\, Z_j.$$

Output and each shift variable appears in linear form and in pairwise interactions with all variables, including itself. A variable's interaction with itself is simply its square, thus leading to the terminology "quadratic."

As does the translog, the quadratic avoids a priori restrictions on elasticities of substitution among the shift variables, or, more importantly, in the general formulation among inputs. But in contrast to the translog, and important for present purposes, it allows for the unconstrained emergence of economies or diseconomies of scale and scope as well as subadditivity[3] The principal disadvantage of the quadratic is that it is not linearly homogeneous in prices. While this violates a theoretical criterion for a well-behaved cost function — namely, that doubling all input prices exactly doubles total costs — the advantages of the quadratic have caused some researchers to prefer it altogether in cost function estimation (Baumol et al., 1982; Roller, 1990; Kaserman and Mayo, 1991). Given the structure of present data and the nature of the questions to be addressed, we do so as well.

4.2 Specifying and Estimating a Net Cost Function

The initial issues to be analyzed concern economies of scale and economies of vertical integration. In order to apply cost theory, it is important to define a utility's output carefully. Clearly, *distribution* is the ultimate output, and we seek a measure of scale economies in that function[4]. *Generation* is the second output dimension with respect to which scale is to be determined and vertical economies (with distribution) are to be measured. Generation differs from the conventional "second" product in a multiproduct cost framework, however, in that it represents an input into the utility's distribution output. In fact, final output — distribution — may originate either with internal generation or with market purchase of power.[5]

To utilize the quadratic cost function, we replace the single variable Q in equation 4.3 with D and G (later designated *DIST* for distribution output and *GEN* for generation). Both now have the quadratic forms that allow for scale effects in each function. In addition, any cost advantage from vertical integration should manifest itself in lower total costs for a utility producing G and D together than those for two stand-alone utilities, each producing the same amount of output as the integrated utility. Such vertical economies may derive from a common indivisible factor or from cost complementarity.

If all fixed costs were common, then a single intercept term α_0 in the equation would suffice. In reality, however, there may be fixed costs specific to distribution, fixed costs specific to generation, as well as fixed and *common* costs. Since only the latter give rise to economies of scope and integration, two additional variables, one for each of the product-specific fixed costs, should be included.[6] The second possible source of economies of integration, cost complementarity, is captured by the interaction term between the two outputs $D \cdot G$. In contrast to the intercept term, cost complementarity focuses on variable cost differences that arise from joint production.

To see the full implications more clearly, we write out the truncated quadratic (this time, with only the two outputs) thus:

$$C = \alpha_0 + \alpha_G + \alpha_D + \alpha_1 D + (\tfrac{1}{2}) \delta_1 D^2 \tag{4.4}$$

$$+ \alpha_2 G + (\tfrac{1}{2}) \delta_2 G^2 + (\tfrac{1}{2}) \delta_{12} D \cdot G.$$

Here α_G and α_D denote the fixed costs specific to generation and distribution, respectively. Positive values for each of these would be expected, leaving α_0 to reflect the differential fixed costs (positive or negative) from joint operation. Regarding the interaction term $D \cdot G$, the sign and magnitude of the coefficient δ_{12} indicate whether the variable costs of producing D and G together are more or less (in each case, by the amount $[\tfrac{1}{2} \delta_{12} D \cdot G]$) than those associated with stand-alone production.

The remaining issue to be resolved for purposes of estimation concerns the "costs" that actually correspond to these outputs. In the present case where G is generation and D is distribution, a perfectly integrated utility incurs and records costs for both "outputs." A stand-alone distributor, however, incurs costs for the distribution function and, by definition, none for generation. Yet, the latter records costs from the purchase of power in the amount D, costs which are passed through to final output price but which do not reflect actual production by the unintegrated utility. The same problem affects measured costs of partially integrated utilities to the extent of their power purchases.

Since the costs of purchased power do not arise from operations of the utility and simply confound measurement of the costs that are[7] the most obvious approach is to net out purchased power costs from the recorded costs of all utilities and to utilize the resulting "value-added" concept of costs. Value added measures those costs corresponding to the amounts of generation and distribution that the utility in fact performs. Value added is employed here in the first set of cost function estimations.[8]

Since the present focus is on firm equilibrium, the relevant costs in this regression are long-run: Total operating expenses include capital costs such as depreciation and interest payments, not simply O & M. Distribution output *DIST* is total mwh sold to residential, commercial, and industrial customers (so-called "final sales," or "sales to ultimate customers") plus sales for resale. Generation output *GEN* is simply total mwh of generation by the utility[9] The fixed-effects term *FCGEN* is included for those utilities that generate at least some power, but since all utilities in the present data set distribute power, an analogous *FCDIST* becomes part of the constant term. All variables are defined in Appendix A, which includes a summary table of variable names.

Column (a) of Table 4.1 reports the results of estimating a truncated form of the cost function, focusing essentially on the output terms shown in text equation 4.4, while column (b) employs a more comprehensive specification. Concerning the former, the R^2 of well over 90 percent and the very high degree of statistical significance on most of the estimated coefficients indicate that the model fits the data quite well. Electric utility costs are positively related to both generation and distribution output, with the quadratic terms *GENSQ* and *DISTSQ* imparting appropriate convexity to the relationships. As would be expected, total costs increase at an increasing rate.

Regarding vertical integration, the most striking finding is the negative and highly significant interaction term between *GEN* and *DIST*. This confirms cost complementarity, that is, lower variable (and hence total) costs for utilities that simultaneously engage in generation and distribution relative to separate production of equal amounts of *GEN* and *DIST*. The other dimension of vertical economies is the possibility of fixed-cost savings. Neither the constant term in this regression nor that designed to capture the fixed costs specific to generation (*FCGEN*), however, is statistically significant. Read literally, the latter indicates positive fixed costs of generation, offset by a term suggesting cost savings from the sum of overall fixed costs plus those attributable to distribution.[10]

These results regarding vertical economies deserve further examination, but before doing so we estimate a more comprehensive cost function. That specified in column (b) of Table 4.1 adds to the above model a number of independent variables for competition, various cooperative and institutional arrangements, type of generating capacity, the utility's transmission and distribution systems, fuel and other costs, and various output characteristics. A brief description of each of these factors is provided here, with full details in the data appendix.

Competition. Utilities operating in anything other than the standard monopoly environment are identified by the fixed effects term *COMP*. *COMP* takes on a value of unity for all the cases identified in Tables 3.2 and 3.3, that is, duopoly, new hookup competition, and border competition.

Cooperative and Institutional Arrangements. Many utilities operate under cooperative or other institutional arrangements that may affect operating performance and costs. Some of these may in fact be alternatives to vertical integration. These include the following:

■Membership in holding companies. A variable *HCSUB* is defined as unity for subsidiaries of registered holding companies. Measured costs of these subsidiaries might differ due to economies of coordination or simply because of accounting practices.

Table 4.1. NET COST EQUATION		
VARIABLE (scale)	(a)	(b)
DIST	3.72 (0.70)	13.17 (2.84)
DISTSQ (10^{-6})	2.53 (7.33)	1.25 (4.18)
GEN	42.6 (7.70)	27.46 (5.67)
GENSQ (10^{-6})	2.24 (6.50)	1.20 (4.16)
DIST·GEN (10^{-6})	-4.73 (6.91)	-2.32 (3.99)
FCGEN (10^{6})	11.4 (0.80)	-25.8 (1.45)
COMP (10^{6})		24.5 (1.04)
DIST·COMP		-8.50 (7.11)
POOLMEM (10^{6})		-34.9 (1.77)
DIST·POOL		6.59 (5.71)
HCSUB (10^{6})		65.4 (1.91)
DIST·HC		-13.97 (8.36)
INCENTR (10^{6})		28.0 (0.83)
DIST·INCENT		3.88 (2.84)
GASUTIL (10^{6})		-3.37 (0.28)
NUCLEAR (10^{6})		204.0 (3.58)
HYDRO (10^{6})		-3.87 (0.14)
OTHER (10^{6})		17.1 (0.66)
HIVOLT (10^{6})		-81.5 (3.03)

Table 4.1. NET COST EQUATION (Continued)		
VARIABLE (scale)	**(a)**	**(b)**
RESSIZE (10^6)		-4.83 (2.49)
COMSIZE (10^6)		-0.25 (1.69)
INDSIZE		34.79 (0.38)
PRFSTE (10^6)		0.87 (1.84)
PRFNUC (10^6)		8.00 (5.57)
WAGE (10^6)		3.07 (0.26)
COSTCAP (10^6)		72.0 (0.48)
REGIONS		*
CONSTANT (10^6)	-1.37 (0.14)	69.9 (0.94)
R^2	0.937	0.963
F	0.326	387
N	543	543

Note: t-statistics in parentheses.
      * indicates region dummies included.

■Power pools. The variable *POOLMEM* is defined as unity for those utilities that belong to an organized power pool. With functions ranging from coordination to central dispatch and investment planning, pools may provide an alternative to true vertical integration in the effort to conserve on costs.

■Incentive regulation. Utilities that have been relieved of the most stringent forms of rate-of-return regulation may show gains in efficiency. Those subject to "inventive regulation" are identified by the variable *INCENTR*.

■Combined gas and electric utilities. The zero-one variable *GASUTIL* identifies those electric utilities that also supply natural gas. Possible economies or diseconomies from such diversification should emerge through this variable.

Type of Generating Capacity. Utilities have different mixes of generating plant technologies, each with distinctive costs. Hydro power is relatively inexpensive, whereas standby capacity for peak use only is costly. When first installed, nuclear power was anticipated to be low-cost, but environmental and safety concerns over the past 20 years have altered its cost characteristics. The percent of hydro, nuclear, and other (largely peaking) capacity are denoted by *HYDRO, NUCLEAR*, and *OTHER*, respectively. Cost effects are thus compared to conventional steam (fossil-fuel) generation.

Fuel and Other Input Prices. For those utilities than generate at least some power, the price at which they purchase fossil fuels for conventional steam and peaking units and the price of any nuclear fuel represent major cost determinants. The variables *PRFSTE* and *PRFNUC* denote these two (hydro power uses no fuel). In addition, for all utilities, hourly wages *WAGE* and the cost of capital *COSTCAP* are included to reflect important differences in the prices of these factors. For pure distribution utilities, labor and capital represent essentially the only input costs.

Output Characteristics. Certain output characteristics affect generation and distribution costs; notable among these is the percent of high voltage power and the size of customers. High-voltage power requires little or no voltage reduction before delivery to customers and therefore entails smaller line losses. *HIVOLT* is measured as power sold into industrial and resale uses. Larger average customer size may affect expenses associated with customer accounts, customer service, sales, and even some parts of A & G. Average sizes of residential, commercial, and industrial customers are separately measured as *RESSIZE*, *COMSIZE*, and *INDSIZE*, respectively.

Regions. Otherwise unmeasured differences in input costs or production conditions are allowed for by including fixed effects variables for nine regions of the country in which utilities operate. These variables are based on standard NERC definitions.

Many of these additional variables could be included in multiple forms — as fixed effects, in quadratic form, and as interactions with one or even both measures of output. In actual practice, this would result in extensive proliferation of very highly correlated variables, many with uncertain justification, unlikely to have separate importance, and obscuring the significance of those that do belong in the model. Present analysis therefore focuses on those of inherent interest and which have some a priori justification.

We begin by noting that the output variables carried over from column (a) generally maintain sign and statistical significance. One exception is the term *FCGEN*, which reverses sign but is not particularly significant in either regression. The other change concerns *DIST*, whose coefficient now rises to statistical significance. Still, costs remain convex in each output, and cost complementarity holds. The full regression model confirms the existence of significant economies of vertical integration. The overall fit of this regression is very strong, with an R^2 of .963 and an extremely large F-statistic of 387.

The magnitudes of most of these coefficients are somewhat reduced in column (b). Indeed, that on the interaction *DIST•GEN* is only half as large as previously. These changes result from the inclusion in column (b) of a number of interaction terms with output (*DIST•HCSUB*, etc.) terms obviously correlated with output and which share some of the quantitative impact of output on costs. It is also worth noting that these latter interactions — those with *HCSUB*, *POOLMEM*, *INCENTR*, and *COMP* — have greater statistical significance than the corresponding fixed effects variables.

Among these other variables, we note first that *COMP* appears with a positive (but insignificant) coefficient, whereas the interaction term *DIST•COMP* is negative and quite significant. Taken literally, the former suggests that competition raises a utility's fixed operating costs by about $24 million, an amount representing 9.9 percent of typical total costs.[11] The coefficient on the interaction between competition and distribution output, however, implies lower variable costs (O & M) for competitive utilities. Evaluated at the means of variables and netting out the change in fixed costs, the net cost reduction for competitive utilities is fully 16 percent.

This number, while strikingly large, is not inconsistent with some earlier findings. It suggests that utility competition involves additional fixed costs (no doubt from such factors as duplicative distribution plant) but lower operating costs thereafter. These results largely persist through the various models to follow and confirm that competition indeed appears capable of reducing the total costs of utility operation by a substantial amount.

A similar phenomenon emerges with respect to subsidiaries of holding companies. Such subsidiaries appear to have greater fixed costs (in this case, by $85 million) but have significantly lower operating costs.[12] For the average size utility, these effects net out to a very large cost savings of 18 percent. As discussed earlier, while such an effect could be due to accounting conventions, it might also reflect real coordination economies achieved through holding company affiliation as an alternative to vertical integration.

It might be noted that controls for two other interfirm arrangements reveal no significant effect on costs. One is the so-called "exempt holding companies," those not subject to PUHCA due to their intrastate operation and generally exhibiting a lesser degree of coordination. The second is "joint action agencies," limited cooperative ventures existing strictly among publicly owned systems. Neither is included in these reported regressions in order to limit the number of variables.

Somewhat different patterns arise in the cases of the other two variables, *POOLMEM* and *INCENTR*. Membership in a power pool appears to lower fixed utility costs by $30 million, consistent with the view that pool membership can in fact achieve cost efficiencies similar to those from vertical integration. The trade-off appears to be that variable utility costs are significantly greater for power pool members. The net effect for the typical utility is a cost *increase* of 5.9 percent.

This finding of higher net costs for pool members replicates that in the literature. Since many utilities were pressured to form or join pools, especially in the aftermath of the 1965 Northeast blackout, it may be that utilities correctly do not anticipate net benefits from pools. An alternative possible explanation is that pools arise where costs are above some norm, but the pools fail to lower costs below that norm. This would produce a spurious association between higher costs and pool membership. Yet another possibility is that cost savings result only from "tight" pools, roughly defined as those that coordinate dispatch. In an unreported regression, a separate control for tight pools fails to confirm that refinement.

The fixed effects variable for IOUs subject to incentive regulation and its interaction with distribution output both emerge with unexpectedly positive signs in this regression. Others have found a similar association and concluded (as with pool membership) that incentive regulation has probably been put into effect for utilities with the most troubled cost structures.[13] It is doubtful that incentive regulation literally causes higher costs than the alternative of rate-of-return. Rather, one probably should conclude simply that the data offer no support for cost savings from incentive regulation.

The last among these "characteristics" variables is *GASUTIL*, which captures any cost differences from combined gas and electric utilities relative to undiversified electricity operation. The small negative value is statistically quite insignificant, indicating the absence of any real effect from combined operation. Alternative specifications, such as the relative proportions of assets in electric versus non-electric operations, also fail to disclose any systematic effects.[14]

Among the lengthy list of control variables, we note that the percentage of generation capacity devoted to nuclear power (*NUCLEAR*) is associated with significantly higher costs, while neither hydro nor other non-steam power generation has much effect either way. In unreported regressions, a measure of the vintage of each utility's capital proved insignificant. These effects persist throughout the remaining regressions.

Other output characteristics affect costs in predictable ways. A larger percent of output that is high-voltage (*HIVOLT*) results in significantly lower costs. This no doubt reflects lower line losses and reduced costs of voltage conversions in industrial and resale segments of the market. For residential and commercial power, greater average usage per customer (*RESSIZE* and *COMSIZE*, respectively) lowers total costs. Both distribution costs and customer service costs are potentially saved in the case of high-volume users. This effect does not hold for larger industrial users (*INDSIZE*), who may already be large enough for most economies to accrue to the utility.

Factor prices are expected to be major influences on total costs. Four factor prices are included in column (b), namely, steam fuel, nuclear fuel, labor, and capital. These all emerge with the correct positive signs, and two of them — the steam and the nuclear fuel prices — are statistically significant. These results confirm that higher factor prices are indeed reflected in increased costs. Finally, fixed effects of nine geographical regions are allowed for. They are often, but not always, collectively significant.

These estimated cost functions clearly establish the relationship of utility costs to generation and distribution outputs and to a variety of other factors. Notable among the latter is the role of competition in reducing variable costs, even as facilities duplication imposes a fixed-cost penalty. While membership in holding companies and in power pools might serve, in principle, as alternatives to vertical integration, the evidence is quite mixed as to their effects on cost components. Incentive regulation and diversification into gas distribution have no beneficial cost effects. Other control variables, on the other hand, perform very much as expected in influencing utility costs. The next two sections take a closer look at vertical and scale economies, followed by a discussion of such economies under public versus private ownership.

4.3 Economies of Vertical Integration and Scale

For further analysis of the effects of vertical integration on utility costs, we apply the necessary and sufficient condition for positive vertical economies. As previously identified in equation 3.1, this is that

$$C(G, D) - C(G, 0) - C(0, D) < 0. \tag{4.5}$$

For the quadratic cost function in equation 4.4, this is equivalent to

$$(\tfrac{1}{2}) \, \delta_{12} D \cdot G - \alpha_o < 0 \tag{4.6}$$

Although it would appear that equation 4.6 could be applied to either of the specifications reported in Table 4.1, two related difficulties arise with respect to the column (b) results. First, the *regression* constant reported there is not the α_o required in equation 4.6, since the regression contains numerous other independent variables that would have to be appropriately aggregated as well. Second and more serious is the fact that some of the right-hand side variables are themselves output-dependent and hence not really constants at all. Terms like *DIST•COMP* vary with distribution output and thus cannot be aggregated at, say, mean values. Under these circumstances, accounting for other terms becomes a yet more complex and statistically less reliable undertaking.

The analysis in this section therefore utilizes the results in column (a) of Table 4.1.[15] In those results, both the constant term α_0 and the coefficient on $D \cdot G$ are negative, so that the condition required for positive economies of vertical integration — equation 4.6 — depends upon their relative magnitudes. It is immediately clear that the inequality is satisfied except when D and G are sufficiently small that the first term of equation 4.6 is (algebraically) greater than α_0. Clearly, then, vertical diseconomies may arise, but only for small-output combinations.

The exact range within which this may occur and the magnitude of economies for a wide range of output combinations can be calculated directly. Specifically, we compute the percent of excess costs due to stand-alone generation and distribution using a variant of the formula in equation 3.4. The variant is given by

$$S_v = \frac{C(G,0) + C(0,D) - C(G,D)}{C(G,0) + C(0,D)} \quad . \tag{4.7}$$

This equation differs from the previous expression for percent cost savings in that its denominator is the sum of stand-alone costs, rather than the costs of joint production.[16]

Panel A of Table 4.2 reports this calculation for the larger output range essentially in two million mwh increments, while panel B examines the small-output "corner" in 200,000 mwh increments. It is evident from this table that all sizeable output combinations realize cost efficiencies from vertical integration. Negative values appear only in the smallest output ranges of panel B, and even then the magnitude of such diseconomies never exceeds seven percent. Clearly, most output combinations benefit from joint production, and these economies grow quite substantial with volumes. In panel A, for example, the cost penalties of stand-alone production rise from six percent for a combination of one million mwh of both generation and distribution, to over 50 percent for combinations above 10 million mwh each. Interestingly, for any given amount of distribution output, the largest (positive) vertical economies arise when generation output is approximately equal.

The average utility in the present data set distributes about five million mwh and is 32 percent integrated, implying about 1.5 million mwh generation. Such a utility has 22 percent lower total costs from integrated operation. The average *output* combination in the data set (as opposed to the average across *utilities*) consists of five million mwh distribution, together with four million mwh generation. Such an output combination achieves an estimated cost savings from integration of 31 percent.[17] The typical IOU has yet larger distribution and generation outputs, and hence realizes apparent vertical economies of 55 to 60

Table 4.2. Vertical Economies

Panel A

		Generation (million mwh)									
		1	2	4	6	8	10	12	14	16	18
	1	0.06	0.07	0.08	0.08	0.07	0.07	0.07	0.06	0.06	0.06
	2	0.11	0.15	0.16	0.15	0.15	0.14	0.13	0.12	0.12	0.11
Distribution	4	0.16	0.23	0.28	0.28	0.27	0.26	0.25	0.24	0.23	0.22
(million mwh)	6	0.16	0.26	0.34	0.37	0.37	0.37	0.35	0.34	0.33	0.32
	8	0.15	0.25	0.37	0.42	0.44	0.44	0.44	0.43	0.42	0.40
	10	0.13	0.24	0.37	0.44	0.48	0.50	0.50	0.50	0.49	0.47
	12	0.12	0.22	0.36	0.45	0.50	0.53	0.54	0.55	0.54	0.53
	14	0.11	0.20	0.35	0.44	0.51	0.55	0.57	0.58	0.58	0.58
	16	0.10	0.19	0.33	0.43	0.50	0.55	0.59	0.60	0.61	0.62
	18	0.09	0.17	0.31	0.41	0.49	0.55	0.59	0.62	0.63	0.64

Panel B

		Generation (million mwh)									
		0.2	0.4	0.6	0.8	1	1.2	1.4	1.6	1.8	2
	0.2	-0.07	-0.04	-0.02	-0.01	-0.01	0.00	0.00	0.00	0.00	0.01
	0.4	-0.05	-0.02	-0.01	0.00	0.01	0.01	0.02	0.02	0.02	0.02
Distribution	0.6	-0.04	-0.01	0.01	0.02	0.03	0.03	0.03	0.04	0.04	0.04
(million mwh)	0.8	-0.03	0.00	0.02	0.03	0.04	0.05	0.05	0.05	0.06	0.06
	1	-0.02	0.02	0.04	0.05	0.06	0.06	0.07	0.07	0.07	0.07
	1.2	-0.01	0.03	0.05	0.06	0.07	0.08	0.08	0.09	0.09	0.09
	1.4	0.00	0.04	0.06	0.07	0.08	0.09	0.10	0.10	0.10	0.11
	1.6	0.00	0.04	0.07	0.08	0.09	0.10	0.11	0.11	0.12	0.12
	1.8	0.01	0.05	0.08	0.09	0.10	0.11	0.12	0.13	0.13	0.13
	2	0.01	0.06	0.09	0.10	0.11	0.12	0.13	0.14	0.14	0.15

percent. By contrast, the average publicly owned system, with just over 900,000 million mwh of distribution and 150,000 generation, faces modest *dis*economies on the order of three percent.

Although vertical economies are only part of the explanation for utility structure, these results are striking in their consistency with actual practice. Small utilities appear to be best structured as deintegrated operations, and indeed they are. For larger scale combinations of output — certainly any utility with more than one million mwh in distribution — the economies of integration are compelling. Virtually all utilities of such size are highly integrated. Indeed, one might say that they can be of such size only by integrating.

Overall optimum firm size and structure are a function of scale economies in generation and in distribution as well as of vertical economies. The reason is that, even in the presence of vertical economies, sufficiently strong diseconomies of scale may outweigh the cost advantages of integration. So-called "stage-specific economies of scale," analogous to product-specific economies in the more typical single-product context, are measured as the ratio of average incremental cost of a particular output to its marginal cost. This captures the extent of cost savings to be gained by "scaling up" that one output.

For distribution output, stage-specific economies are represented as:

$$S_D = AIC_D / MC_D \qquad (4.8a)$$

$$= ([C(G,D) - C(G,0)]/D) / MC_D . \qquad (4.8b)$$

Values greater than one indicate positive economies, since marginal cost is less than average incremental cost. For the quadratic cost function estimated here, this equals

$$S_D = \frac{\alpha_D + \alpha_1 D + (\tfrac{1}{2}) \delta_1 D^2 + (\tfrac{1}{2}) \delta_{12} D{\cdot}G}{\alpha_1 D + \delta_1 D^2 + (\tfrac{1}{2}) \delta_{12} D{\cdot}G} \qquad (4.9)$$

Calculations of S_D and an equivalent expression S_G for generation prove to be quite sensitive to chosen output combinations, the result of a large cross-product term between generation and distribution.[18] For representative values of each, however, estimated stage-specific economies are reasonable. The "average" utility with five million mwh of distribution and 1.5 million mwh of generation has an S_D of 1.43; and the average output combination of five million mwh distribution and four million mwh generation has S_D of 1.68. Both indicate substantial economies of distribution out through the range of typical operation. Indeed, this is also the case for the average size IOU, whose S_D is 2.66, and for the average publicly owned utility, whose small size results in extremely large estimated economies of distribution.

With respect to generation, the calculated values of S_G indicate that these economies are exhausted well within the range of operation of these utilities. In fact, only the typical publicly owned utility, with distribution of 900,000 mwh and generation of 200,000 mwh, has a value in excess of one ($S_G = 1.24$). For the average utility overall, for the average output combination, and for the average IOU, the values are 0.85, 0.85, and 0.47, respectively. All these imply the absence of strong product-specific economies in generation, much as other studies have shown.

The final step of this analysis synthesizes economies of vertical integration with generation and distribution economies into a measure of "multistage economies of scale." This captures the *overall* degree of scale economies in distribution plus generation, allowing for trade-offs between the two. Thus, favorable economies of vertical integration might offset diseconomies of each individual stage, justifying larger firm sizes, if integrated, than implied by stage-specific analysis. Alternatively, vertical diseconomies might erode positive economies in distribution and in generation, resulting in deintegrated (but perhaps large) firms. The conventional measure of multistage economies is given by

$$S_M = \frac{C(G,D)}{G \cdot MC_G + D \cdot MC_D} \tag{4.10}$$

Algebraic manipulation of equation 4.10 reveals that S_M is a weighted average of the degrees of scale economies in G and D, adjusted for economies of integration between the two (see Bailey and Friedlaender, 1982). Values in excess of unity imply overall multistage economies.

The calculated values of S_M in Table 4.3 are once again quite striking in their ability to explain actual patterns of firm structure. Panel B indicates significant multistage economies for the entire range of small output combinations. For the very smallest, this is presumably the result of strong generation and distribution economies offsetting diseconomies from integration. For somewhat larger sizes, integration ceases to be a disadvantage, and stage-specific economies continue to dominate.

Most of the largest output combinations in panel A display no such systematic benefit of multistage operation. Except for a narrow band down the main diagonal of the table, most points indicate multistage *dis*economies. That band represents "balanced" integration, in the sense that G is approximately equal to D. Here the weakening or reversal of scale economies in generation and

Table 4.3. Multistage Economies

Panel A

		Generation (million mwh)									
		1	2	4	6	8	10	12	14	16	18
	1	1.22	1.09	0.96	0.88	0.83	0.79	0.75	0.73	0.71	0.69
	2	1.13	1.11	1.01	0.93	0.87	0.82	0.78	0.75	0.73	0.71
Distribution	4	0.87	0.99	1.05	1.01	0.95	0.89	0.84	0.80	0.77	0.75
(million mwh)	6	0.72	0.83	0.98	1.03	1.01	0.96	0.91	0.86	0.82	0.79
	8	0.64	0.72	0.88	0.98	1.02	1.01	0.97	0.92	0.88	0.84
	10	0.60	0.66	0.78	0.90	0.98	1.01	1.00	0.97	0.93	0.89
	12	0.58	0.62	0.71	0.81	0.91	0.98	1.01	1.00	0.98	0.94
	14	0.56	0.59	0.66	0.75	0.84	0.92	0.98	1.00	1.00	0.98
	16	0.55	0.57	0.63	0.69	0.77	0.85	0.93	0.98	1.00	1.00
	18	0.54	0.56	0.60	0.66	0.72	0.79	0.87	0.93	0.97	1.00

Panel B

		Generation (million mwh)									
		0.2	0.4	0.6	0.8	1	1.2	1.4	1.6	1.8	2
	0.2	2.08	1.55	1.36	1.25	1.19	1.14	1.10	1.07	1.05	1.03
	0.4	1.97	1.54	1.37	1.27	1.20	1.16	1.12	1.09	1.07	1.05
Distribution	0.6	1.83	1.51	1.36	1.27	1.21	1.17	1.13	1.11	1.08	1.06
(million mwh)	0.8	1.68	1.46	1.34	1.27	1.22	1.18	1.15	1.12	1.10	1.07
	1	1.54	1.40	1.31	1.26	1.22	1.18	1.15	1.13	1.11	1.09
	1.2	1.41	1.33	1.28	1.24	1.21	1.18	1.15	1.13	1.11	1.10
	1.4	1.30	1.27	1.24	1.22	1.19	1.17	1.15	1.14	1.12	1.10
	1.6	1.20	1.20	1.20	1.19	1.17	1.16	1.15	1.13	1.12	1.11
	1.8	1.12	1.14	1.15	1.16	1.15	1.15	1.14	1.13	1.12	1.11
	2	1.05	1.09	1.11	1.12	1.13	1.13	1.13	1.12	1.11	1.11

distribution is offset by significant economies of vertical integration between the two.[19] Larger firm sizes are now cost-justified, so long as they are more or less fully integrated. Deintegration or substantially unbalanced integration, however, results in a cost penalty.

The importance of balanced integration is evident from inspection of the regression results in Table 4.1. Setting $G = D$ implies that the quadratic terms for generation and distribution, which represent the convexity of costs, can be combined with the cross-product term that represents (positive) vertical economies. Stated in terms of a single output denoted by Q, the cost function for the strictly integrated firm would therefore equal

$$C = 46.3 \ Q + .04 \ (10^{-6}) \ Q^2 + 10.0 \ (10^6). \tag{4.11}$$

The coefficient on Q^2 in this equation — .04 — results from summing those on *GENSQ* and *DISTSQ* (4.77) and netting out that on *DIST·GEN* (-4.73). This small but positive coefficient implies that balanced integration leads to costs that rise almost linearly with output. The cost function is not strictly linear, however: An F-test of the hypothesis that the sum of the coefficients on *GENSQ*, *DISTSQ*, and *DIST·GEN* equals zero results in $F(1,536) = 2.56$, significant at no better than 11 percent. Thus, there is a slight nonlinearity, sufficient eventually to cause diseconomies to set in and thereby to limit firm sizes.

In actual practice, utilities adopt structures consistent with these results. Large utilities are on average much more highly integrated than are small utilities. The top half of the size distribution generates 55 percent of its distribution requirements, as compared to only nine percent for the lower half. The simple correlation between distribution output and the percent of own-generation is a highly significant 0.53. In general, large utilities need to integrate to achieve cost efficiencies, but, for small utilities, balanced integration is less important. Distribution and generation outputs, in short, are supplied by utilities structured in terms of size and integration that reflect available economies.

4.4 Specifying and Estimating a Total Cost Function

Economies of scale and integration represent perhaps the most complex of the major cost issues to be examined. Together with certain related matters, those have been the focus of the preceding sections. Here attention turns to the impact of public versus private ownership on costs. We begin, however, by returning to a data issue that determines how this and subsequent examination of costs must proceed.

Derivation of net cost requires data on purchased power expenses. A number of utilities, however, routinely report as "purchased power expenses" amounts that represent the *difference* between their payments for purchased power and the revenues received from sale of power to other utilities. Since only some of the errors can be identified, net cost data are subject to some reliability questions.[20] Those data have nonetheless been employed for purposes of analyzing questions of scale and integration, since those issues are directly linked to generation and distribution outputs. For our remaining purposes, however, a different approach is adopted, one that is less subject to such error and, fortunately, not fundamentally different where its implications can be compared.

The alternative approach involves use of *total* recorded costs but includes purchased power as a control variable. The latter should effectively adjust costs for varying amounts of power actually purchased by each utility and thus permit measurement of the impact of other variables on implicitly corrected total cost. Purchased power should appear with a coefficient that represents average purchase price to utilities.[21] This alternative approach has the further merit of similarity to the various empirical models estimated throughout the literature and used for comparison purpose in this study.

We begin by demonstrating the approximate equivalence of results based upon total cost controlling for power purchases, with those based on net costs. Columns (a) and (b) of Table 4.4 replicate the regression models in Table 4.1 but with total costs as the dependent variable and power purchases as a control variable. All of the estimated coefficients have signs and significance levels quite similar to those in Table 4.1. Magnitudes may differ, of course, since the dependent variable is now total cost rather than net cost.

In this regression, the coefficient on *PURCH* itself is positive, highly significant, and readily interpreted as the price of purchased power. The value of 3.67 cents in column (a) is very close to the average price of purchased power of 3.9 cents reported in Chapter 2, confirming the role played by this variable. Clearly, this approach does a good job of correcting the data for the costs of purchased power without dramatically altering the essential findings based on net costs.[22] We now wish to use this model to explore the impact of public ownership on economies of scale and vertical integration.

To this end, column (c) of Table 4.4 adds to the model in column (b) a group of variables allowing coefficients on the output terms to differ between publicly owned and investor-owned utilities. *PUBLIC* is a fixed-effects variable for publicly owned systems, while *PUB•DIST*, *PUB•DISTSQ*, *PUB•GEN*, *PUB•GENSQ*, *PUB•D•G*, *PUB•DIST*, and *PUB•PURCH* represent the interactions between *PUBLIC* and the corresponding output term.[23] Each estimated coefficient should capture any difference due to public ownership in the cost impact of the corresponding output term.

As is immediately clear, most of these additional variables fall well short of standard levels of statistical significance. Only the power purchase interaction *PUB•PURCH* comes close, and among the true output terms, the largest t-value appears on *PUB•D•G*. Collectively, these additional variables do not achieve statistical significance either. An F-test on the joint explanatory effect of all the public ownership terms results in $F(7,499) = 1.32$, significant at no better than 24 percent. Thus, there is no real evidence that publicly owned utilities have a significantly different cost function than do privately owned systems, after controlling for all other variables. They differ, in short, in what they do but not in how they do it.

That said, it remains of interest whether any single one or few such terms might be statistically significant and economically important. Even if all are not significant collectively, those that are would imply ownership differences in particular facets of the distribution, generation, and/or vertical functions of the utilities.[24] To determine this, we proceed in a heuristic stepwise fashion. First, *PUB•PURCH*, which has the largest t-value of the public ownership variables, is added to the regression in column (b), followed by *PUB•DIST*, with the second highest t-statistic. Other variables are examined for their incremental explanatory power, ultimately resulting in the column (d) regression.

Column (d) contains the four public ownership variables that come closest to individual statistical significance and collectively contribute significant explanatory power: *PUB•DIST, PUB•GEN, PUB•D•G,* and *PUB•PURCH*. None has a t-value less than 1.54 (implying significance at 12 percent or better), and the F-test for their joint significance yields $F(3,503) = 2.20$, which is significant at eight percent. While not definitive, these variables would appear to be doing a good job of capturing what differences do exist between publicly owned utilities and IOUs.

Read literally, the estimated coefficients offer potentially important insights. First, as suggested by the crude data comparison, it appears that publicly owned utilities actually pay more *at the margin* for purchased power than do IOUs. Second, publicly owned systems appear to have a cost advantage in the distribution function relative to IOUs. Third and by contrast, IOUs seem to achieve lower costs than do publicly owned systems in power generation. Fourth, economies of vertical integration, so important for utilities generally, are significantly smaller for publicly owned utilities. The coefficient on *PUB•D•G* suggests that cost complementarity for the latter is nearly 25 percent smaller.

Collectively, these results suggest that public and private utilities have comparative advantages in *different facets* of the electric utility industry. Public ownership achieves lower costs in the end-user function, consistent with earlier speculation that their perspective and motivation might lead them to perform better

VARIABLE (Scale)	(a)	(b)	(c)	(d)
Table 4.4. TOTAL COST EQUATION				
DIST	7.77 (1.02)	38.32 (5.73)	39.42 (5.70)	39.31 (5.75)
DISTSQ (10^{-6})	2.74 (6.66)	2.09 (6.03)	2.25 (6.23)	2.18 (6.18)
GEN	37.7 (5.04)	2.78 (0.42)	0.64 (0.09)	1.36 (0.20)
GENSQ (10^{-6})	2.30 (5.53)	1.90 (5.47)	2.07 (5.85)	1.99 (5.72)
DIST·GEN (10^{-6})	-4.98 (6.05)	-3.81 (5.54)	-4.12 (5.84)	-3.98 (5.74)
FCGEN (10^{6})	10.4 (0.70)	-26.5 (1.43)	30.7 (0.61)	-19.1 (1.01)
PURCH	3.67 (4.61)	5.63 (0.80)	3.40 (0.46)	3.30 (0.45)
COMP (10^{6})		26.4 (1.07)	29.4 (1.17)	30.5 (1.23)
DIST·COMP		-10.41 (8.33)	-10.50 (8.40)	-10.5 (8.43)
POOLMEM (10^{6})		-30.1 (1.45)	-34.8 (1.64)	-33.8 (1.63)
DIST·POOL		6.01 (4.83)	5.93 (4.70)	5.96 (4.79)
HCSUB (10^{6})		84.6 (2.34)	87.4 (2.31)	83.7 (2.28)
DIST·HC		-14.28 (7.91)	-14.3 (7.18)	-14.2 (7.86)
INCENTR (10^{6})		33.8 (0.96)	33.9 (0.93)	33.0 (0.93)
DIST·INC		4.71 (3.25)	4.97 (3.30)	4.93 (3.41)
GASUTIL (10^{6})		-8.74 (0.71)	-7.21 (0.58)	-7.13 (0.58)
NUCLEAR (10^{6})		205.0 (3.44)	203.0 (3.32)	213.0 (3.54)
HYDRO (10^{6})		-5.89 (0.20)	-12.5 (0.41)	-8.49 (0.29)
OTHER (10^{6})		8.88 (0.33)	7.51 (0.28)	6.82 (0.25)
HIVOLT (10^{6})		-80.4 (2.87)	-65.0 (2.25)	-63.4 (2.20)
RESSIZE (10^{6})		-6.95 (3.43)	-6.09 (2.90)	-5.99 (2.90)
COMSIZE (10^{6})		-0.26 (1.69)	-0.202 (1.30)	-0.209 (1.36)

Table 4.4. TOTAL COST EQUATION (Continued)				
VARIABLE (Scale)	(a)	(b)	(c)	(d)
INDSIZE		32.1 (0.33)	25.5 (0.26)	27.6 (0.29)
PRFSTE (10^6)		0.83 (1.68)	0.723 (1.46)	0.784 (1.58)
PRFNUC (10^6)		8.65 (5.77)	8.65 (5.17)	8.55 (5.13)
WAGE (10^6)		-3.38 (0.27)	-3.59 (0.28)	-2.76 (0.22)
COSTCAP (10^6)		64.4 (0.41)	86.1 (0.54)	83.6 (0.53)
PUB·DIST			-90.2 (1.60)	-102.4 (1.88)
PUB·DISTSQ			-21.1 (0.99)	
PUB·GEN			72.4 (1.33)	81.7 (1.56)
PUB·GENSQ			-33.4 (0.85)	
PUB·D·G (10^{-6})			6.18 (1.39)	0.90 (1.54)
PUB·FCG (10^6)			-54.4 (1.06)	
PUB·PURCH			10.0 (1.94)	9.81 (1.90)
PUBLIC (10^6)			40.3 (0.80)	
REGIONS		*	*	*
CONSTANT (10^6)	-0.851 (0.08)	131.0 (1.69)	72.0 (0.78)	105.0 (1.34)
R^2	0.948	0.969	0.970	0.970
F	1383	456	373	412
N	543	543	543	543

Notes: t-statistices in parentheses.
        * Indicates dummy variables included.

in local distribution. No advantage accrues to publicly owned systems in the upstream generation task. Rather IOUs realize greater efficiency in power production and from vertical integration as well. Both integration of investor-owned generators into more local distribution and integration of publicly owned distribution utilities into generation extends each system into tasks for which the other is better suited. The evidence shows that on balance IOUs gain more from such integration than do public systems.

These findings differ from most previous analyses that have come (or have attempted to come) to judgments about the overall superiority of one type of utility. Present evidence suggests that such efforts are mistaken. Neither public ownership nor private ownership demonstrates superiority in all tasks within the electric power industry. Rather, each mode has a distinctive strength: Small locally owned enterprises perform best in end-user tasks such as power distribution, while larger and privately owned utilities achieve superior performance where this local orientation is less important.[25]

Striking corroboration of this conclusion may be found in actual experience. As displayed in Table 4.5, 59 percent of publicly owned utilities are distribution-only, concentrating on the very activity for which they have a cost advantage and avoiding generation where they do not. By contrast, only five percent of IOUs are distribution-only. Virtually all of the latter both generate as well as distribute power, reflecting their advantage in generation as well as vertical economies that they appear more adept in capturing. A chi-square test of significance of the frequency pattern in Table 4.5 results in a test statistic of 123.8, overwhelmingly significant and rejecting any notion that this distribution is due to chance. Public systems, in short,are unintegrated distribution specialists for a reason.

We conclude this part of the analysis by quantifying the effects of public ownership on costs. Based on the regression in column (d) of Table 4.4 and calculated at the mean values of all variables, the net effect of public ownership is to reduce a utility's costs (relative to IOU costs) by 5.5 percent. This is the result of about an average 11 percent reduction in distribution costs, offset by increases of 4.4 percent in generation, 0.5 percent in vertical integration, and 0.6 percent in power purchases. Obviously, the magnitude of cost savings in specific cases depends on output volumes and mix.

These results imply that the source of the apparent cost advantage associated with public ownership is more likely to be found in end-user orientation than in regulation or subsidies. Subsidies have been fully accounted for and a substantial cost differential remains. If regulation were the culprit, then publicly owned utilities should realize cost efficiencies in both generation and distribution. The fact that the cost advantage of public ownership is located in the distribution task and that investor-owned utilities have comparable cost advantages in generation is compelling evidence that each has distinctive strengths.

In summary, we find that, after controlling for outputs, factor prices, and numerous other variables, utility costs are significantly affected by vertical integration, by public ownership, and by competition. All three institutions on average reduce total costs, although in each case the effect has important variations. Vertical economies do not obtain for certain small utilities. Public ownership has

SIZE	OWNERSHIP		
	PRIVATE	PUBLIC	TOTAL
Distribution Only	8	233	241
Generation Plus Distribution	139	163	302
TOTAL	147	396	543

Table 4.5. FREQUENCY DISTRIBUTION OF UTILITIES

different effects on distribution versus generation costs and on vertical economies. Competition appears to increase fixed costs, but to lower variable costs by more. Thus, simple statements about the form of these effects are as unwarranted as outright dismissal of them altogether.

4.5 Cost Structure of a Large Public System: The Case of Los Angeles

The finding that the cost consequences of ownership differ according to the task being examined raises a number of intriguing questions. Among these is whether large utilities, regardless of whether they are investor-owned or publicly owned, are structured similarly and incur similar costs. That is, do large public systems have more in common with large IOUs than with other smaller publicly owned utilities? If such is the case, the imperatives of large institutions must dominate other forces affecting firm structure and cost performance.[26]

One piece of evidence on this question derives from a case study of the largest publicly owned electric utility in this country, the Los Angeles Department of Water and Power. A so-called "diagnostic audit" was commissioned by the Los Angeles City Council in November 1993 and conducted by an outside consulting firm. Its ostensible purpose was to investigate whether the Department of Water and Power (DWP) was minimizing costs at a time when imminent changes in the electric power industry would be making cost efficiency increasingly important.

The resulting May 1994 report (Barrington – Wellesley, 1994) is largely based on public data supplied to FERC. It offers a large number of simple comparisons of costs, staffing, and rates between DWP and three benchmark panels:

■Urban IOUs (eight large utilities from around the country)

■California electric utilities (the three California IOUs plus the Sacramento Municipal Utilities District (SMUD))

■Publicly owned utilities (seven medium to large public systems — SMUD plus six from other states)

Table 4.6. TOTAL COST EQUATION

VARIABLE (Scale)	(a)	(b)	(c)
DIST	36.5 (5.07)	39.4 (5.77)	3.14 (0.50)
DISTSQ (10^4)	2.06 (5.60)	2.06 (5.88)	0.026 (0.10)
GEN	5.38 (0.76)	1.22 (0.18)	-12.1 (2.71)
GENSQ (10^6)	1.94 (5.32)	1.91 (5.47)	-0.860 (3.19)
DIST GEN (10^4)	-3.81 (5.28)	-3.78 (5.47)	0.76 (1.47)
FCGEN (10^6)	3.08 (0.22)	-23.8 (1.26)	-2.31 (0.19)
PURCH	13.9 (1.90)	6.87 (0.96)	22.2 (4.48)
PUBLIC (10^6)	2.73 (0.11)	0.222 (0.01)	-0.196 (1.25)
PUB DIST	-18.1 (2.74)	-7.06 (1.07)	7.77 (1.79)
PUB D G (10^6)	1.65 (3.84)	0.67 (1.34)	0.56 (1.97)
COMP (10^6)	44.9 (1.72)	27.8 (1.11)	-2.39 (0.14)
COMP DIST	-10.8 (8.19)	-10.4 (8.35)	-3.48 (3.90)
HCSUB (10^6)	44.6 (1.16)	77.5 (2.06)	26.5 (1.05)
HC DIST	-14.0 (7.24)	-14.0 (7.61)	-11.2 (7.74)
POOLMEM (10^6)	21.7 (1.18)	-29.8 (1.41)	-14.5 (1.06)
POOL DIST	4.97 (3.92)	5.87 (4.66)	6.84 (7.41)
INCENT (10^6)	108. (2.91)	33.1 (0.90)	34.7 (1.42)
INCENT DIST	3.70 (2.37)	4.82 (3.21)	1.59 (1.53)
GASDUM (10^6)		7.08 (0.57)	-0.94 (0.12)
CAPNUC (10^6)		218. (3.61)	165 (3.98)
CAPHYD (10^6)		-6.61 (0.22)	-29.3 (1.51)
CAPOTH (10^6)		8.38	3.67

Table 4.6 - TOTAL COST EQUATION (Continued)

VARIABLE (Scale)	(a)	(b)	(c)
PCTHIVO		-76.5 (2.70)	-53.1 (2.88)
AVGMRES (10^6)		-6.39 (3.08)	-4.33 (3.22)
AVGMCOM (10^6)		-0.24 (1.55)	-0.158 (1.58)
AVGMIND		25.6 (0.26)	-3.62 (0.06)
PRFSTE (10^6)		0.801 (1.61)	-0.191 (0.59)
PRFNUC (10^6)		7.90 (4.82)	-8.69 (3.97)
WAGE (10^6)		-1.01 (0.08)	-8.75 (1.02)
COSTCAP (10^6)		75.7 (0.47)	-105 (0.98)
PRFS GEN			1.70 (22.8)
PRFN GEN			1.01 (8.67)
WAGE DIST			0.615 (0.96)
CCAP DIST			273. (7.06)
REGIONS		*	*
CONSTANT (10^6)	0.33 (0.01)	110. (1.33)	163. (2.93)
R^2	0.963	0.969	0.987
F	766	420	922
N	543	543	543

Notes: t-statistics in parentheses
* Indicates dummy variables included

The report finds that DWP's average electric costs exceed those of all three panels.[27] Further analysis, here summarized in Table 4.6, identifies A & G, O & M, and distribution costs as relatively high. DWP's customer service expenses are comparable to the benchmark panels, while its transmission expenses are lower. Average salary and wages are considerably below that of Urban IOUs or other California Utilities. Much of this appears to be offset, however, by lower productivity (mwh sold) per employee. That, the report concludes, reflects differential staffing by DWP. As reproduced in Table 4.7, DWP staffing levels exceed, seemingly by a wide margin, the IOU benchmark average for A & G, T & D, and electricity supply. Only in the customer service area is its staffing less than the norm.

The report does not provide analogous data with respect to electricity supply expenses, but it states that DWP's production expenses per mwh are lower than those of Urban IOUs and California Utilities and higher than those of the Public Systems panel. Its purchased power expenses are higher than all

Table 4.7. STAFFING LEVELS BY FUNCTION		
ITEM	DWP TOTAL	IOU AVERAGE
Administrative & General	3,194	1,839
Customer Service	1,033	1,033
Transmission and Distribution	2,923	2,389
Electric Supply	1,818	1,083
TOTAL ELECTRIC	8,968	6,360

Source: Barrington–Wellesley (1994)

comparison panels, due to unfavorable long-term contracts with major suppliers.

In some respects, Los Angeles's experience follows that of other public systems, whereas in others it more closely resembles any large utility regardless of ownership. As reviewed in Table 3.4, publicly owned utilities generally have higher O & M, distribution, and purchased power expenses, and lower wage costs, all of which hold for Los Angeles as well. But its relatively low generation costs and higher A & G, customer service, and customer accounts expenses more closely resemble those of large, integrated, investor-owned utilities. While the picture is mixed, it is clear that Los Angeles differs from the "typical" public system.

This latter observation is confirmed by inspection of the data in Table 4.6. For virtually all items, DWP levels are closer to Urban IOUs than to other Public Systems. As suggested before, it may simply be the case that larger entities take on certain common characteristics, including more substantial bureaucracies and heavier staffing levels, regardless of whether they are privately or publicly owned. The imperatives of large organizations — the external imperative of developing and tending its customer bases, the internal imperative of control in large organizations — may simply dominate any potential for public systems to conserve on costs. This would help to explain the convergence of costs between IOUs and large public systems and the tendency for public systems to be the smaller utilities.

4.6 Summary

The analysis conducted in this chapter has established a number of important results regarding utility costs and structure. It has shown, for example, that vertical integration achieves significant cost efficiencies, in some circumstances sufficient to offset diseconomies of scale in generation and distribution separately. Public ownership also proves capable of lowering costs, but that effect arises only in the distribution phase. By contrast, investor-owned utilities have advantages in generation and, on balance, in vertically integrating. Large integrated utilities appear to have similar cost structures regardless of their mode of ownership.

These are central findings of the present cost study. Each adds important insight and refinements to our understanding of the effects of ownership, integration, and competition on utility costs. The complex nature of the benefits of vertical integration, the selective role of competition, and, perhaps most especially, the distinctive impact of public ownership are all revealed in this analysis. We next turn to the implications of cost and other influences on electric utility pricing.

Notes

1. A good review of these is provided in Berndt (1991).

2. Certain partial remedies for this problem are possible. For example, zero-one variables may be respecified in exponential form, and other data may be transformed using the Box – Cox method. The resulting estimation is more cumbersome, and there is little reason to prefer this aw kward alternative.

3. The inability of the translog to contend with zero output values effectively precludes a finding of economies of scope.

4. In this study, "distribution" refers to transmission plus distribution. In this we follow virtually all other studies. Further disaggregation proves unwieldy.

5. We return below to some implications of this fact for the measurement of costs.

6. More technically, product-specific fixed cost terms permit U-shaped average incremental costs for each product. In the actual empirical specification employed below, only one such term — that for generation — is included since all utilities engage in distribution.

7. One possible caveat is that cost conservation may manifest itself in purchased power expenses as well.

8. Compatibility with subsequent modeling, together with certain errors in the net cost data, ultimately leads to an examination of total cost data as well. This will be discussed in the next section.

9. On average, final sales represent 87 percent of total distribution output. Some generated or purchased power does not appear in distribution. This consists of line losses, energy furnished without charge, and energy used by the company itself.

10. The negative sum of overall fixed costs plus those from distribution is somewhat implausible. This result is probably best interpreted simply as inconsistent with the proposition that vertical integration confers benefits via fixed-costs savings.

11. It represents 12.1 percent of *net* costs, but total costs are the more familiar metric

12. One might expect the opposite in the case of holding company subsidiaries. That is, by holding reserve capacity jointly, conservation of fixed costs should result. The exact reasons for this pattern remain unclear.

13. Berg and Jeong (1991) report that incentive regulation techniques were introduced in states where firms performed suboptimally, but that such alternative regulation did not really improve cost performance. See also St. Marie (1996).

14. Both Mayo (1984) and Sing (1987) have examined combination gas and electric

utilities. Neither found evidence of positive economies of scope.

15. It should also be noted that the results in column (b), as well as others later in this chapter, are quite consistent with those in column (a). Reliance upon this set of results imparts no known bias to the conclusions. Present analysis resembles that of Kaserman and Mayo, although there are major differences as well: use of net rather than total cost, far larger and more diverse sample of utilities, etc.

16. Use of stand-alone costs regularizes the percentage calculation. Joint costs vary quite widely, especially in the extreme ranges of output, making it a less useful standardizing factor.

17. The standard formula used to compute Table 4.2 takes the cost difference as a percent of the cost of *integrated* operation. Relative to the cost of stand-alone operation, this represents a 27 percent cost savings from integration.

18. The large estimated coefficient on *DIST·GEN* leads to calculated average incremental costs and marginal costs thatvary widely, and by virtue of 4.8(a), to S_G and S_D that can be unreliable in extreme ranges.

19. Recall that the largest vertical economies arose in the region of similar G and D in Table 4.1.

20. This problem even results in some utilities reporting *negative* purchased power expenses. Censoring those variables at zero introduces unknown biases in other observations where identical misreporting occurs but is not evident.

21. There is some misreporting of power purchases, but much less than for purchased power expenses. The former are recorded in an account where netting out is less likely, and indeed only two utilities report negative purchases.

22. More specifically, these results imply similar scale and vertical economies. As before, balanced integration leads to very modest diseconomies associated with larger size.

23. Rather than testing a production cost relationship, *PUB·PURCH* tests whether publicly owned and privately owned utilities pay the same amount per unit of purchased power. This helps resolve the question raised in Chapter 2.

24. Such a possibility is in fact suggested by the t-value on the coefficient on *PUB·D·G*, as well, perhaps, as those on other terms.

25. Publicly owned utilities may resemble local retailers whose specialized knowledge of and connections to their markets allow them to compete successfully with larger and seemingly more efficient competitors. Market- and even customer-specific knowledge may confer a cost or quality advantage on such sellers.

26. In an analogous context, Lynk (p. 458) observes that "a nonprofit hospital organization whose only function is the provision of hospital services to a well-defined population, and whose governing board effectively represents that same population, looks — and probably acts — a lot more like a consumer cooperative that a creator of monopoly resource misallocation. But, in contrast, the same nonprofit hospital when controlled by those with an interest in funding other unrelated activities, may look indistinguishable from a for-profit operation".

27. It is worth noting that the report's benchmark panels are comprised of relatively small numbers of utilities and lack controls for the various ways in which their operations differ. In addition, since Los Angeles constitutes the largest publicly owned system in the country, comparisons are inevitably to smaller utilities.

Chapter 5

PRICE AND MARKUP BEHAVIOR

The price of electric power is perhaps the most visible criterion by which utility performance is judged. It is more obvious in the market than costs and it matters more to consumers. Those facts alone would justify attention to price and its determinants. In addition, while costs prove to be the most important influence on price, the relationship between costs and price is significantly affected by differences in ownership, competition, and other features of the market and the utility itself. For these reasons this chapter examines price and pricing behavior by electric utilities.

That costs are the primary determinant of price is not surprising, but it is important to note that price may *diverge* from costs for several reasons. One possibility previously discussed is that the decision maker, that is, the utility owner, manager, or regulator/administrator, may choose not to pass cost differences through to consumers. Instead, cost savings might be used for some other objective, or if cost inefficiencies arise, price may or may not fully reflect such overruns.

One specific source of inefficiency is rate-of-return regulation, which almost certainly results in excess total costs and perhaps excess utilization of capital in particular. The former effect, which may be termed "cost-inefficiency,"[1] arises from the absence of competitive pressures on producers, as is the case with most publicly owned and investor-owned electric power companies. Excess capitalization (the so-called "Averch – Johnson effect") serves to increase the rate base and therefore the profitability of a utility subject to rate of return regulation in particular, since allowed profit is calculated on that rate base.[2] Publicly owned enterprises are subject to cost inefficiencies, too, as their product or production process may be used to gain political support for management or the sponsoring politicians. Support may derive from hiring excess workers, paying excess wages, or entering into contracts on terms lucrative to outside parties, all actions that elevate costs above necessary levels.

These observations imply that while the price of electricity is critically influenced by costs, it is by no means completely determined by them. It is this further relationship of price to ownership and competition that represents the focus of the present chapter. We first review theory and evidence regarding utility prices and then statistically analyze the various determinants of electric power pricing.

5.1 Theory and Evidence on Utility Pricing

Any reasonable theory of electric utility pricing must begin with costs. However, precisely how costs affect price depends in turn upon the objectives of regulated, competitive, or publicly owned utilities, and on the process that governs price determination. Here we briefly review four simple models of the relationship of price[3] to underlying costs.

First, electric utilities could be argued to price at marginal cost since that maximizes economic efficiency. Assuming for simplicity that marginal cost is constant, Figure 5.1 displays such costs as MC and labels the resulting price P_c. Yet with high fixed costs, marginal-cost pricing results in the failure of utilities to cover their full costs of operation. Again, in Figure 5.1, if average total costs are denoted by the curve AC, note that AC is everywhere above MC by the amount of average fixed costs. The only remedy for the revenue shortfall would appear to be direct subsidization of such an entity, a phenomenon that few believe occurs.[4] Simple marginal-cost pricing is not thought likely to govern utility pricing.

Second, some studies have concluded that prices under rate-of-return regulation differ little, if at all, from profit-maximizing levels. By analogy, it could be argued that public systems are perhaps managed in order to maximize profits and hence revenue contributions to municipalities (although there is no evidence of such net flows). To that extent, pricing under both of those regimes might be modeled as simple profit-maximization. Such pricing would require marginal cost to be equated to marginal revenue, resulting in the price P_m in Figure 5.1. Despite some mixed evidence, the view that utilities succeed in maximizing profitability, that is, that they operate without any effective pricing constraint, is almost certainly too extreme.

A more plausible variant of this model would claim that utilities do mark up price over marginal cost, but not to the full profit-maximizing level. The operative question then becomes: What determines where price is set between marginal cost and profit-maximizing levels? One possibility for the private monopoly firm is that it simply extracts whatever rents it can, given oversight by a regulator who is neither omniscient nor completely ineffectual. For the publicly owned firm, excess revenues could be sought for transfer into the municipal general fund for local tax reduction or expenditure increases. While quantitatively imprecise, this objective of generating "some" excess revenues merits attention.

A more precise variant of this scenario represents the fourth and perhaps most plausible theory of pricing: The regulator or public administrator might

Figure 5.1

Alternative Prices for Electric Utilities

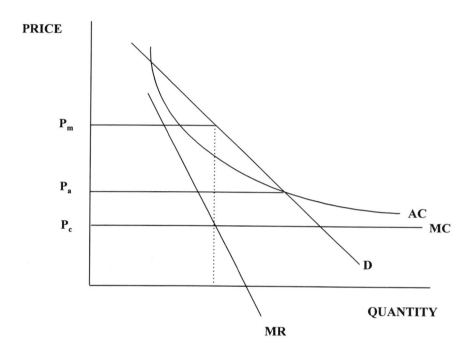

explicitly or implicitly sanction excess revenue extraction sufficient to cover fixed costs, so that the markup is essentially a device to permit the utility — whether regulated or publicly owned — to break even. More specifically, recovering full costs of operation implies that price must equal the sum of marginal cost plus per unit fixed costs, which is, of course, average total costs. P_a in Figure 5.1 denotes this price. This will constitute our point of departure in modeling price setting for most utilities.

This diversity of price-cost models is mirrored in the many comparisons of publicly owned versus investor-owned electric utilities. The most extensive early theoretical treatment of pricing under public ownership is due to Peltzman (1971), who models public ownership as a mechanism by which municipal tax revenues are used to subsidize prices to favored customer groups. In order to maximize political support for the utility, its managers, and/or municipal politicians, Peltzman predicts lower price for publicly supplied power, the greatest benefit for the modal customer group, less differentiation ("discrimination") in price schedules under public ownership, and a looser relationship between price and cost. He offers data that support some of these propositions and attempts a complex but ultimately crude effort to control for costs.[5]

A more sophisticated approach to the cost basis of pricing is taken by Moore (1970). In order to test the effectiveness of regulation and of public ownership in restraining prices, Moore estimates long-run marginal costs and demand elasticities. He then compares calculated profit-maximizing price for each utility with its actual price. Moore concludes that regulation at best reduces price by five percent, whereas public ownership achieves a 10 to 22 percent price reduction relative to unconstrained profit maximization.

Other early studies are less convincing. DiLorenzo and Robinson's (1982) analysis comes to the same conclusion as Moore, although their treatment of costs is rudimentary. DeAlessi's (1975) study of wholesale power (i.e., resale) finds lower prices from publicly owned utilities, but those data may simply reflect very low-cost federal hydro power projects rather than subsidization.[6] Mann and Seifried (1972) test whether costs matter at all in price determination by publicly owned utilities, but since few argue that *only* political factors are relevant, this demonstration does not seem particularly informative.

Recent analyses based on explicit cost functions permit more precise tests of pricing, and especially, price-cost relationships, for public and private utilities. Two studies in particular deserve note. Hayashi, Sevier, and Trapani (1985) estimate marginal costs and demand functions for residential, commercial, and industrial segments of the market. They conclude that price-cost margins for IOUs fell from breakeven levels[7] in the 1960s to near marginal cost levels by 1980. Public systems' margins on residential and commercial service lay between these extremes (also declining during this period), while industrial power appears to have been priced at, or possibly below, marginal cost, an unexpected result under public ownership. Hollas, Stansell, and Claggett (1994) find lower prices for residential and commercial users of public power relative to rural coops, but find seemingly higher prices for industrial users. Their approach controls for a number of other factors, including unit costs of operation by the utility.

Far less extensive work has been conducted on the price implications of competition among electric utilities. Hellman's previously cited work concluded that competition promoted lower prices, but this was based on fairly crude comparisons. Primaeux's work is somewhat more systematic and constitutes the major body of literature on this subject. He finds large (as much as 33 percent) and significant reductions in most residential prices from duopoly versus monopoly municipal firms, after controlling for differences in costs and other factors (Primeaux, 1985). The ability of competitive utilities to price so much lower is attributed to the elimination of the cost inefficiency that Primeaux's other work analyzes.

Apart from public ownership and competition, several other possible influences on utility pricing have been investigated. Joskow (1982) offers empirical support for the proposition that characteristics of utility commissions themselves may affect prices. Primeaux and Mann (1986) hypothesize in particular that *elected* public utility officials are more responsive to politically powerful constituencies, such as residential customers. They nonetheless fail to find any systematic differences in prices between states that elect commissioners and others that do not. Harris and Navarro (1983) critique Primeaux and Mann's work, but their own study also fails to find any such effect.

At least one study has examined the public finance implications of public ownership. Nelson (1980) tests whether revenue contributions by publicly owned utilities to their municipalities result in higher costs and, implicitly, higher prices. He finds that such contributions in fact occur where the utility is operating at *lower* cost, suggesting a different causal chain: Municipalities may take advantage of systems that are *more* efficient by extracting rents in circumstances that do not force prices above some norm.

Although this latter is an interesting observation about public ownership, the model developed below will focus on the more conventional role of costs, ownership, and competition in price determination.[8]

5.2 A Model of Average Cost Pricing

The need for any electric utility, whether publicly or privately owned, to avoid substantial over-earning or under-earning implies that pricing must reflect its costs. The actual process of price determination, which focuses on meeting the utility's "revenue requirements," is consistent with this perspective. For these reasons, average cost pricing may be taken as a good first approximation to utility pricing and an appropriate beginning point for analysis. In fact, we shall see that, properly framed, average cost pricing may even be a fairly complete explanation.

Based on preceding discussion, the average cost pricing model can be written as follows:

$$PRICE = f(AVGCOST, TAXRATE, PUBLIC, COMP, X). \qquad (5.1)$$

The first two right-hand-side variables are cost terms, *AVGCOST* denoting average total costs of the utility, while *TAXRATE* is its per unit rate of tax and tax-like payments. Both constitute portions of the utility's revenue requirements and as such are major determinants of the price level under both regulation and public ownership. The fixed effects variable *PUBLIC* is then included to determine whether, given those costs, price is set differently under public versus private ownership. Similarly, *COMP*, a variable defined as unity for duopoly, new hookup, or border competition, is intended to capture the impact on price from a utility's operation in a competitive environment.

Three other factors that influence overall price are represented in equation 5.1 by the vector *X*. As in the cost function, the dummy variable *INCENTR* identifies those IOUs subject to some kind of incentive regulation instead of traditional rate of return. To the extent that incentive regulation restrains pricing relative to costs, this variable should measure that difference. *ELECTED* denotes IOUs operating in states where public utility commissioners are elected. This will capture any tendency for popular election to lead to lower power price than when rates are regulated by an appointed commission. Lastly, since industrial and resale power both costs less and commands a lower price, the average price of power sold by a utility will be sensitive to its mix of services. *HIVOLT* measures the percent of industrial and resale power.

The results of this regression analysis are reported in Table 5.1. We begin in column (a) with a truncated version of this model, focusing on the two cost terms plus *PUBLIC*, *COMP*, and the control for output mix *HIVOLT*.[9] The estimated coefficients on all variables have the expected signs and are statistically significant. Those on average cost (*AVGCOST*) and tax rate (*TAXRATE*) make clear that the price of electric power is overwhelmingly determined by these cost factors. Their magnitudes would be expected to equal unity in a regime of complete pass-through of costs to the price of electric power, but both fall short of unity by at least a modest margin. *HIVOLT* confirms the importance of output mix in the average price of power.

The coefficient on *PUBLIC* is also statistically significant, indicating that publicly owned utilities do indeed price power lower than IOUs. The magnitude of the coefficient — 0.15 cents per kwh — represents a 2.5 percent decrease relative to average price for all utilities. It should be stressed that this price-reducing effect of public ownership is distinct from any cost effect, since the present estimate holds costs (and output mix) constant; that is, for a given cost level, the price under public ownership is 2.5 percent less. It is also noteworthy that the estimated price and cost effects together fully explain the gross (i.e., unadjusted) price differential of 7.4 percent noted in Chapter 2. Public ownership has previously been estimated as responsible for a 5.5 percent cost savings[10] and now for a 2.5 percent price differential given those costs, for a total of eight

Table 5.1. PRICE EQUATION				
VARIABLE	**(a)**	**(b)**	**(c)**	**(d)**
AVGCOST	.872 (52.6)	0.868 (51.6)	0.870 (51.9)	0.873 (52.1)
TAXRATE	.700 (11.8)	0.690 (11.5)	0.691 (11.6)	0.681 (11.5)
PUBLIC	-0.0015 (2.46)	-0.0013 (1.70)	-0.0014 (1.72)	-0.0012 (1.57)
COMP	-0.0020 (2.08)	-0.0021 (2.24)		
HIVOLT	-0.0071 (5.80)	-0.0071 (5.78)	-0.0070 (5.68)	-0.0070 (5.72)
ELECTED		-0.0008 (0.64)	-0.0008 (0.67)	-0.0008 (0.67)
INCENTR		0.0014 (1.26)	0.0013 (1.19)	0.0011 (1.00)
DUOPOLY			-0.0048 (3.04)	-0.0009 (0.36)
HOOKUP			0.0050 (1.67)	
BORDER			-0.0011 (0.95)	
PUB·DUOP				-0.0063 (1.96)
CONSTANT	0.0122 (9.10)	0.0122 (8.62)	0.0121 (8.51)	0.0119 (8.39)
R^2	0.874	0.875	0.876	0.876
F	746	534	419	473
N	543	543	543	543

Note: t-statistics in parentheses.

percent.

The effect of competition on price is indicated by the coefficient on *COMP*. The magnitude of 0.20 cents per kwh represents a 3.4 percent decrease in average price by competitive utilities and is statistically significant. This result confirms the standard economic prediction of competition as a pricing discipline. Together with the previously estimated effect of competition on average costs (a 16 percent reduction), this implies a total competitive effect on cost plus price of nearly 20 percent. The precise role of competition will be examined further below. Overall this regression and all others in this table fit the data very well, with R^2s in excess of .87.

The regression in column (b) incorporates two other possibly relevant variables — namely, *INCENTR* and *ELECTED*. As it did in the cost regression, *INCENTR* appears with an unexpected positive sign, but is insignificant. Here, too, this may simply indicate that incentive regulation is put into place where utility pricing conduct (like its cost performance) is particularly poor, not that incentive regulation actually makes matters worse.[11] Popular election of state public utility commissioners does carry the expected negative sign for its price effect, but in this and the remaining results in this table, the coefficient on *ELECTED* is well short of statistical significance.[12]

The final regressions in columns (c) and (d) explore the role of competition more thoroughly. In reality, the *COMP* variable encompasses three types of "competitive" utilities — true duopolies, new hookup regimes, and competition at the boundary of service territories. In place of the single dummy variable, the regression equation (c) distinguishes these three. Strikingly, duopoly competition, indicated by the variable *DUOPOLY*, is negative and significant, whereas neither weaker form of competition — *HOOKUP* and *BORDER* — have such an impact. Hookup competition appears with a positive sign, and border competition cases show no significant effect at all. It would appear that only head-to-head competition makes a difference, although it is should be noted that there are only a few utilities in such market circumstances. The estimated coefficient on *DUOPOLY*,.0048 per kwh, implies a very sizeable 8.1 percent price effect.

This effect is an average across both IOUs and publicly owned systems. Regulatory constraints on large IOUs, however, might result in a smaller measured effect for IOU duopolists than for the publicly owned. The reason is that IOU service territories are typically much larger than the one community where they may face a duopoly competitor. Since their price structure is generally constrained to be the same, or at least not much different, throughout their service territory, any response by the IOU to a lower price charged by a small publicly owned utility in one community is likely to be muted.

Column (d) therefore designates public systems in duopoly markets by *PUB•DUOP*, leaving IOU effects to be captured by *DUOPOLY* itself. Both prove to be negative, but only *PUB•DUOP* now approaches statistical significance.[13] It would appear to be publicly owned systems in duopoly markets that are responsible for the most substantial price reductions. The estimated effect for them is fully 10.6 percent. While duopoly IOU prices are lower, they are lower by only about 1.5 percent, thus corroborating the belief about their constrained response.[14]

In summary, it seems quite clear that both public ownership and duopoly competition are associated with lower prices than those for conventional regulated monopoly IOUs. These effects hold true even after possible cost differences have been accounted for. Together with the results in the preceding chapter, these imply that public ownership succeeds in lowering costs, and then after doing so, in lowering prices further. The effect of competition similarly involves cost reduction, followed by a reduction in price relative to those costs. The next section casts further light on these results, based on an alternative model of the relationship between utility price and cost.

5.3 Markup Pricing by Utilities

The major alternative model of utility pricing is based on a linkage between price and *marginal*, rather than average, cost. As previously noted, the degree of price markup may reflect simply an effort to extract rents, or it may be intended only to cover fixed costs. This section describes the relationship of price to marginal cost and then sets out an approach to estimation that avoids some difficulties of the traditional formulation. The following section reports the estimation, compares the findings with those based on average cost pricing, and assesses some significant limitations of this new approach.

    A profit-maximizing firm chooses output so that its marginal costs equals its *perceived* marginal revenue. Its perceived marginal revenue is the usual marginal revenue derived from market demand, adjusted for the individual firm's degree of pricing discretion relative to that of a pure monopolist (who faces the entire market demand).[15] This can be formalized as follows: Letting $D = f(P)$ represent market demand (equivalent to distribution output), the pure monopolist's markup is given by

$$MR = P - D/(-f_p), \tag{5.2}$$

where f_p is the derivative of $f(P)$. The expression in 5.2 is thus maximum markup for the stated demand and cost conditions, and implies the price P_m in Figure 5.1.

    For a firm facing a rival in the market (that is, an oligopolist), the operative marginal revenue is no longer MR, but rather it is market MR adjusted for the degree of competition that the rival provides. For example, the rival might respond to a price increase by the firm in question by matching the increase. Or the rival might hold its own price constant, thereby reducing the oligopolist's previous demand. This latter reduction in the oligopolist's pricing discretion is a measure of competition or rivalry between firms. In the present case of electric utilities, the binding external constraint on pricing discretion is that imposed by the regulator (in the case of IOUs) or the public authority (for publicly owned systems). In essence, the utility is assumed to strive for profit-maximization but to be restrained by regulation or by public ownership. It is therefore the difference in the restraint imposed by regulation versus public ownership that is of interest here.[16]

With or without a constraint on pricing discretion, direct calculation of markup generally requires an estimate of marginal cost. The difficulties of accurate marginal cost estimation have led to development of a technique for inferring markups and differences in markups that does not require actual marginal costs. In this approach, it is only the *determinants* of marginal cost that must be represented. If any of those determinants change, then marginal cost, profit-maximizing price, and, ultimately, markups change in predictable ways. Conversely, knowledge of an actual price change resulting from an observed change in the determinants of marginal cost, or in other exogenous conditions such as ownership differences, allows the degree of market power to be inferred.

This last observation is the basis for the present analysis. The exact details can be set out in a simple model of the market in which a monopoly utility operates.[17] First, the demand function that it faces, the market demand, can be written explicitly as:

$$D = \alpha_0 + \alpha_1 P + \alpha_2 Y. \tag{5.3}$$

Here D and P denote market quantity and price, respectively, and Y is an exogenous demand shifter ("income"). From this can be derived an expression for marginal revenue, analogous to equation 5.2).

Next, marginal cost must be expressed in terms of its determinants, which include distribution output D, generation G, a vector of factor prices W, and exogenous supply shifters Z. These imply the following form for marginal cost:

$$MC = \beta_0 + \beta_1 D + \beta_2 G + \beta_3 W + \beta_4 Z. \tag{5.4}$$

Equating this to the derived marginal revenue expression and rewriting gives the monopoly seller's "quasi-supply" function[18] as:

$$P = [\beta_0 + \beta_1 D + \beta_2 G + \beta_3 W + \beta_4 Z] - Q/(-\alpha_1). \tag{5.5}$$

This expression effectively describes the seller's markup over marginal cost in terms of the demand and marginal revenue conditions which it faces.

The constrained monopolist's *perceived* marginal revenue differs from that in equation 5.5, since it no longer chooses output and price off the entire demand curve. Rather, it faces a portion of demand that depends upon the intensity of competition, or, in the present case, regulatory or ownership constraints. We therefore define a parameter θ to represent the degree of pricing discretion available to the firm. The parameter θ ranges from zero to unity as pricing goes from completely constrained (that is, competition) to unconstrained monopoly.

In the above model, θ multiplies the term $Q/(-\alpha_1)$ that represents the difference between marginal cost and price, so that the firm's quasi-supply function is now as follows:

$$P = [\beta_0 + \beta_1 D + \beta_2 G + \beta_3 W + \beta_4 Z] - \theta D/(-\alpha_1). \tag{5.6}$$

Comparing this expression to the preceding makes clear that θ indexes the degree to which the firm can set price between the extremes represented by competition ($\theta = 0$) and monopoly ($\theta = 1$).

This quasi-supply relationship and the demand function in equation 5.3 represent a simultaneous equation system in Q and P that can be estimated econometrically. Although this model does not yet distinguish pricing discretion under public versus private ownership or in duopoly versus monopoly markets, the necessary modification is straightforward. Instead of specifying θ to be identical across all observations, as is the case in equation 5.6, we write

$$\theta = \theta_0 + \theta_1 PUBLIC + \theta_2 DUOPOLY. \tag{5.7}$$

Now θ_1 represents the difference in markup behavior for publicly owned systems and θ_2 the difference for duopoly systems, all relative to monopoly IOUs. The baseline markup for the latter is captured in the term θ_0.

Substituting equation 5.7 into equation 5.6 and combining terms, we obtain the following estimating equation:

$$P = \beta_0 + \beta_2 G + \beta_3 W + \beta_4 Z - [\beta_1 + \theta_0/(-\alpha_1)] D$$
$$+ [\theta_1/(-\alpha_1)] PUB{\cdot}D + [\theta_2/(-\alpha_1)] DUOP{\cdot}D. \tag{5.8}$$

The coefficient on D is expected to be positive, reflecting the quasi-supply response of IOUs and utilities generally. To the extent that public ownership constrains pricing beyond that for benchmark utilities, however, the bracketed coefficient on $PUB{\cdot}D$ will be negative. Analogously, a negative coefficient on the bracketed coefficient on $DUOP{\cdot}D$ would imply a tighter pricing constraint on duopoly utilities. In addition, the values of those coefficients can be used to infer the actual magnitudes of the ownership and competition effects. Since α_1 is estimated simultaneously in the demand function, θ_1 and θ_2 can be recovered from the full model by multiplying the estimated coefficients on $PUB{\cdot}D$ and $DUOP{\cdot}D$ by α_1.

The result is a direct measure of the difference in pricing discretion.[19]

5.4 Evidence on Ownership, Competition, and Markups

The markup model is a simultaneous equations model consisting of equations 5.3 and 5.8 with appropriate variables included for the vectors of demand shifters Y, factor prices W, and supply shifters Z. Earlier cost function estimation included numerous factor prices and supply shifters, but not all of those are relevant to the determination of *marginal* cost. For example, quadratic output terms and interactions with prices are eliminated in the process of taking the quantity derivative of total costs. In addition, certain cost shifters and factor prices — notably, that for capital — do not affect marginal costs. These considerations result in a more parsimonious regression specification.

Table 5.2 reports estimation results for the quasi-supply function in text equation 5.8.[20] This equation is estimated using instrumental variables, with the elements of the vectors Y, W, and Z as instruments. Column (a) focuses strictly on price as a function of distribution output and the interactions of output with dummy variables for public ownership and for duopoly competition. This simple specification nonetheless captures much of interest. The positive and significant coefficient on *DIST* (D in equation 5.8) confirms that price is an increasing function of distribution output, as would be expected from theory. The interaction term *PUB·DIST* measures the difference in price required to elicit a given quantity from publicly owned utilities relative to IOUs. Its negative and significant coefficient implies that publicly owned utilities indeed charge a lower price than do IOUs, as found previously.

Also consistent with previous findings is the estimated effect of duopoly competition. The coefficient on the variable *DIST·DUOP* is negative, although its t-value of 1.59 falls a bit short of conventional significance levels. It is nonetheless suggestive of a price-reducing effect of duopoly.

A substantially better overall fit to the regression is obtained by including an appropriate set of control variables, as shown in column (b). R^2 more than doubles, and while the magnitudes of the key coefficients change, their signs do not, and their statistical significance persists except for some further weakening of the duopoly variable. The coefficient on *PUB·DIST* in column (b) now suggests that power from publicly owned utilities is on average priced 0.217 cents per kwh or 3.7 percent lower. This differential is moderately larger than the 2.5 percent differential estimated in the single-equation, average-cost pricing model. This version adds support to the conclusion that publicly owned utilities indeed price power relative to costs differently than do IOUs.

Similarly, the estimated coefficient on *DIST·DUOP* is negative, implying

Table 5.2. MARKUP EQUATION		
VARIABLE (Scale)	**(a)**	**(b)**
DIST (10^{-6})	0.21 (2.89)	1.02 (2.63)
PUB·DIST (10^{-6})	-4.14 (4.00)	-2.36 (2.85)
DUOP·DIST (10^{-6})	-0.56 (1.12)	-0.27 (0.67)
GEN (10^{-6})		-1.19 (2.96)
HCSUB		-10.9 (3.92)
POOLMEM		10.7 (6.33)
INCENTR		5.27 (1.81)
NUCLEAR		3.42 (0.53)
HYDRO		-9.90 (3.18)
OTHER		7.80 (3.05)
PRFSTE		0.083 (1.91)
PRFNUC		6.73 (3.03)
WAGE		-3.02 (3.23)
HIVOLT		-22.3 (7.93)
CONSTANT	61.2 (59.4)	80.8 (16.8)
N	543	543
R^2	0.150	0.319
F	6.89	20.0

Note: t-statistics in parentheses.

lower prices from duopoly systems relative to conventional monopoly utilities. Its magnitude implies a 0.253 cent differential, or 4.3 percent, although this coefficient falls well short of conventional significance levels. As in the previous model, this result is nonetheless consistent with a price-reducing effect, primarily due to duopoly competition.

Other variables in column (b) perform according to expectations. The negative and significant coefficient on *GEN* indicates that for a given sales level, marginal cost and thus price are lower for utilities that generate more of their requirements. This is consistent with the finding of cost savings from vertical integration analyzed at length earlier. Also in accord with earlier results, holding companies achieve lower costs which translate into lower prices. As before, pools and incentive regulation appear to arise in circumstances where costs and prices are *above* the norm, probably accounting for their positive coefficients in this regression.

Reliance upon nuclear capacity raises costs slightly, while hydro power significantly lowers costs and price, and "other" or peaking power substantially raises them. Among input costs, only the prices of steam and nuclear fuel are positive and significant or nearly so. The impact of wages is strongly and unexpectedly negative for reasons that are not completely clear.[21] Finally, in keeping with past findings, greater concentration on high-voltage power is associated with lower overall costs and prices.

The complete model, of course, consists of this quasi-supply function together with a demand function. In addition, the latter is necessary in order to deduce the values of θ_1 and θ_2, the differences in markups between IOUs and publicly owned utilities and between monopoly and duopoly utilities, respectively. There is an extensive literature on demand functions to draw upon for the present formulation.[22] The exogenous variables suggested by both theory and past empirical work include the number of customers (here broken down into residential, commercial, and industrial customers); household disposable income; the price of the principal substitute fuel, namely, natural gas (again, by customer category); and climate conditions, specifically, cooling and heating degree days.[23] In addition, since the quantity variable includes wholesale power, here taken as exogenous, the magnitude of sales for resale is included in order to correctly measure the impact of the other explanators on retail sales.

The results in Table 5.3 show that price is negatively and significantly related to output. The magnitude of the estimated coefficient implies a demand elasticity at mean price and quantity of .59, well within the range of estimates reported in the literature. Among other variables, greater numbers of residential and industrial customers significantly increase demand. The unexpected negative sign on commercial customers is probably due to high collinearity among these

Table 5.3. DEMAND EQUATION	
VARIABLE (Scale)	**(a)**
PRICE (10^6)	-47.8 (2.59)
RESCUST	29.3 (17.3)
COMCUST	-57.2 (4.33)
INDCUST	467 (7.48)
INCOMPC	42.0 (0.54)
PRGASR (10^4)	-21.4 (0.77)
PRGASC (10^4)	76.9 (1.71)
PRGASI (10^4)	-48.0 (1.86)
HEATDAY	157 (1.45)
COOLDAY	956 (2.48)
RESALE	1.58 (17.9)
CONSTANT	-40.3 (0.23)
N	543
R^2	0.944
F	819

Note: t-statistics in parentheses.

three variables. As in other studies, greater household per capita income does not appear to significantly increase electricity consumption. Regarding natural gas prices, only that for commercial gas price is positively related to demand. The high correlation of these variables across customer classes undoubtedly contributes to their weak explanatory power in this regression. Greater heating and cooling degree days raise demand, both at or near conventional significance levels. And the correction for resale sales is highly significant.

As previously discussed, the parameter for the difference between pricing behavior under public ownership and that by IOUs, θ_1, is calculated by multiplying the estimated coefficient on *PUB•DIST* in the quasi-supply relation by the estimated coefficient on *PRICE* in the demand function. The results of that procedure for the specification in column (b) of Table 5.2 and Table 5.3 (properly scaled) imply that θ_1 equals 0.113, that is, that the effect of public ownership is to reduce the markup by 11.3 percentage points. As noted, the absolute value of the markup cannot be deduced from this model, but this differential would be consistent, for example, with an IOU markup of 30 percent and a public system markup of about 19 percent.

Comparable calculations can be made for duopoly competition, despite the questionable statistical significance of the estimates. Taken literally, the implied θ_2 that measures the competitive effect on pricing behavior relative to costs equals 1.3 percentage points. This estimated effect is a good deal smaller than that for public ownership, consistent in that respect with the somewhat uncertain impact on price previously found for duopoly or other competition.

Despite the insight provided by these results (and their plausibility), the markup model will not form the basis for further analysis. The reasons are twofold. First, the underlying assumptions of this model are subject to some reservations in the present context. The model is rooted in profit-maximization and presumes that observed markups reflect pursuit of profits subject to competitive constraints. But neither regulated IOUs nor publicly owned utilities are motivated by simple profit-maximization. Rather, both are governed by institutions oriented towards covering full costs and therefore set price at no less than average cost but not necessarily much more.

The second concern with the markup model is that, as written, it presumes that all utilities operate on the same cost curve, and then represents differences in prices given such costs. But the last chapter has found cost differences to be quite important. While the model can be modified to allow for possibly different cost structures,[24] it is not well suited for capturing simultaneous cost and price effects.

For these reasons,[25] the markup model of pricing will not be pursued in the remaining analysis. While it is useful to observe the consistency of its implications with those of the full-cost model, the latter will constitute the basis for further examination.

5.5 Summary

This chapter has modeled electric utility pricing in two quite different ways: first, as average-cost pricing by the regulator or public manager, and second, as markup pricing over marginal costs. The best evidence from these models confirms that publicly owned utilities price power more cheaply than to IOUs, after holding costs and other factors constant. The differential appears to be about 2.5 percent. The effect of competition is statistically less clear, although all the evidence points to a lower price at least from true duopolies and especially those that are publicly owned.

Average-cost pricing and markup pricing represent the most plausible alternative models of price setting in this industry. Of the two, however, average-cost pricing is the more convincing and will be used in the further analysis of pricing performance.

Notes

1. Often termed "X-inefficiency," the phenomenon of cost inefficiency in noncompetitive markets was first examined by Leibenstein (1966). See also Frantz (1992).

2. See Averch and Johnson (1962) and, for a review of the empirical evidence, Berg and Tschirhart (1988).

3. Here we focus on a single price. Issues of price *structure*, that is, prices to different segments of the market, will be addressed later.

4. Interestingly, the claimed subsidization of publicly owned utilities could be viewed as an example of this objective. Yet, as previously discussed, the evidence actually shows a net transfer *from* utilities to the municipality.

5. Among the indications of limitations of his analysis is the finding that the excess of private over public price is greater for *non*residential users. It should also be noted that Peltzman does not check the validity of the assumption that publicly owned utilities are net beneficiaries of revenues flows from municipalities.

6. In addition, it is not made clear how "subsidized" wholesale prices, benefiting nonlocal buyers in the resale market, are supposed to create political advantage to the utility.

7. With multiple customer classes (discussed more extensively in Chapter 6 below), their breakeven price is not simply marginal cost plus average fixed cost. Rather, these are "second-best prices," marked up over marginal cost in inverse proportion to demand elasticity for each customer class. See, e.g., Berg and Tschirhart.

8. It also will exclude vertical integration and certain other factors considered in the cost analysis. The reason is simply the absence of a good a priori basis for believing, for example, that given costs, the price charged by a vertically integrated utility will differ from a nonintegrated one.

9. It should be noted that this regression takes average cost as exogenous, an assumption examined in the next chapter.

10. This is the net cost figure for distribution, generation, and vertical effects relative to IOUs.

11. It might indicate that lower costs from incentive regulation resulted in some residual uncaptured profitability to the utility, but earlier evidence indicated no such cost savings.

12. Inclusion of a Salomon Brothers' rating of state regulatory environments was not significant nor did it improve the explanatory power of *ELECTED* or other variables. For its use in a related context, see Joskow, Rose, and Wolfram (1996).

13. One might say that it is statistically significant in a one-tail test. In any event, *DUOP* and *PUB·DUOP* are jointly highly significant, with an F-statistic of $F(2,534) = 6.05$. This is to be expected, given the significance of the single variable DUOP in column (c).

14. The estimated effect for IOU pricing could indicate either a 1.5 percent reduction

throughout its territory, or where allowed, a larger price reduction in the duopoly area that averages with unchanged prices elsewhere to a system-wide 1.5 percent reduction. Present data cannot distinguish between these possibilities.

15. This model is described and illustrated in Bresnahan (1982, 1988) and in Baker (1990), from which this discussion is adapted. It is now widely used to estimate pricing power.

16. Some possible reservations about this interpretation will be discussed below.

17. Although the model does not require the demand function to be linear, that assumption is convenient for expositional purposes here and ultimately represents the estimating form.

18. This is a *quasi*-supply function, since a monopolist does not have a true (i.e., single-valued) supply function.

19. The degree of pricing discretion for *benchmark* utilities — monopoly IOUs — cannot be recovered, since θ_0 is part of a coefficient on D that includes β_1 as well as α_1. The former is not estimated in this model.

20. The associated demand equation will be discussed later. Instruments from that equation consist of numbers of customers (residential, commercial, and industrial), per capita income, price of natural gas (residential, commercial, industrial), and heating and cooling degree days.

21. As noted in the Data Appendix, wage data are from statewide manufacturing wage compilations due to the very incomplete wage data by individual utility. Where such data have been used before, wages have often not played their expected role. See Hayashi et al.(1987).

22. Benchmark surveys are Taylor (1975) and Bohi (1981). More recent studies will be surveyed in Chapter 6.

23. Definitions of these additional variables and data sources are provided in the Data Appendix.

24. For example, a dummy variable for public ownership can be interacted with various cost determinants. While this may allow for cost differences across ownership modes, it results in a much more complex estimating equation that precludes recovery of the key pricing parameter θ_1.

25. There are other limitations of the markup model as well. See, for example, Boyer (1996).

Chapter 6

EXTENSIONS TO SIMULTANEOUS EQUATIONS AND CUSTOMER CLASSES

Thus far the analysis has involved separate inquiries into the determinants of utility costs and the determinants of the overall price of power. This approach has yielded insights into the roles played by vertical integration, public ownership, and competition, among other factors, and those will remain key conclusions. This chapter generalizes the previous analysis in two significant directions. The first involves a simultaneous-equations model of price, quantity, and cost determination. The second disaggregates demand and pricing into residential, commercial, and industrial customer classes.

While conventional in the literature, the single-equation models of cost and price employed thus far may incompletely characterize causation. For example, in the supply relation average cost is treated as exogenous, even though separate analysis shows that cost is itself a function of output quantity as well as of other factors. Similarly, the basic price equation assumes that output is exogenous, despite the fact that a demand function relates quantity to price (as shown in the markup model). Because of these interdependencies among price, quantity, and costs, a simultaneous equations approach may be appropriate. Indeed, to the degree that interdependencies are empirically significant, the failure to specify a simultaneous equations model runs some risk of mismeasuring the impacts of ownership, competition, and vertical integration.

Regarding customer classes, there are at least two reasons for examining residential, commercial, and industrial customer classes separately. First, the overall average price charged by any utility will reflect its particular service mix — the percent of electric power that is supplied to each class — as well as the prices set for the classes. Since the latter do differ (see Table 2.7), it is important to establish whether or how much of the difference in average price may be due to service mix as opposed to some other factors. Thus far this possibility has been addressed by including a measure of high-voltage power supplied, but that represents less complete allowance for mix effects than disaggregation into customer classes.

A second reason for this closer examination of customer classes is the possibility that public ownership, regulation, and competition may simply have different effects in each class. In fact, one conventional interpretation of public ownership is that it represents a mechanism by which a politically powerful interest group pursues some financial advantage such as a lower price, higher wages, etc. Observed price differences favoring residential customers would seem at least broadly consistent with this view, since they arguably have the greatest political influence. Of course, regulation may also work to the particular advantage of certain customers — perhaps residential, perhaps industrial, depending on their differential ability to exercise political influence in the rate-setting process. In addition, competition may result in benefits to certain classes, specifically those that otherwise are vulnerable to price elevation.

This chapter begins by developing and estimating a simultaneous-equations model of price, quantity, and cost determination, based closely on the extensive work that has preceded. Details inevitably differ, but the results are quite consistent with what has preceded and provide insight into the complex interdependencies of these variables. Next, this chapter disaggregates the model into residential, commercial, and industrial customer classes. The results confirm various of the above hypotheses concerning the differential benefits of public ownership and competition.

6.1 A Simultaneous Equations Model

Each of the major components of the simultaneous-equations model remains quite similar to that which has been set out separately before. Price is represented as the response of the regulator/public controller to cost, cost captures the consequences of the utility's choice of quantity, and quantity is modeled as consumer response to price. What differs is that price, cost, and quantity are now all endogenous, so that a different econometric technique is required.

Here we begin by restating the basic equations, focusing on the roles for ownership, competition, and vertical integration. There are three behavioral equations in the full model. As before, total costs C are represented by a quadratic cost function of the form

$$
\begin{aligned}
C = {} & \alpha_0 + \alpha_1 D + (\tfrac{1}{2}) \delta_1 D^2 + \alpha_2 G + (\tfrac{1}{2}) \delta_2 G^2 \\
& + (\tfrac{1}{2}) \delta_{12} D{\cdot}G + \tau_1 PUBLIC + \tau_2 COMP \\
& + \Sigma_i^m \beta_i W_i + \Sigma_j^n \sigma_i Z_i
\end{aligned}
\tag{6.1}
$$

The principal difference from the previous formulation of total costs in equation 4.4 is that distribution output D, taken together with its square and interaction with generation, are now endogenously determined. Distribution output is identical to total consumption as determined by consumer demand:

$$D = \alpha_0 + \alpha_1 P + \alpha_2 Y. \tag{6.2}$$

While this demand function has also been previously utilized, the present version differs in that price is endogenous. Next, price is set by the regulator in the case of IOUs or the public authority in the case of publicly owned utilities, but in either case is based on major cost items:

$$P = f(AVGCOST, TAXRATE, PUBLIC, COMP, X). \tag{6.3}$$

An identity states the obvious relationship between average and total cost and thus closes the model:

$$AVGCOST = C(.)/D(.). \tag{6.4}$$

An important feature of this model is that public ownership and competition are incorporated into both the price and cost equations simultaneously. This allows their effects on price and on cost to be disentangled with greater precision than using single-equation techniques. Estimation of this three-equation model employs instrumental variable techniques.[1] Results for the cost function are reported in Table 6.1, for the demand function in Table 6.2, and for pricing in Table 6.3.

The cost function estimated here includes all the control variables used in the regressions reported in Table 4.8 in some form. Certain output-related variables that would cause overwhelming endogeneity and collinearity problems are, however, suppressed in the present context. Specifically, these are the output interactions with pool membership, incentive regulation, and holding companies.[2] Column (a) of Table 6.1 includes the simple fixed effects terms for those characteristics, as well as one for utilities facing competition and another for those that are publicly owned. As a check on the stability of results, column (b) adds the four interaction terms with output previously found important for public ownership, together with one for competitive utilities. We add these back also because ownership and competition are major concerns of this research.

That said, in most important respects the present results are quite consistent with those obtained earlier by ordinary least squares. In both regressions reported in Table 6.1, the stand-alone costs of distribution and of generation are

Table 6.1. COST EQUATION		
VARIABLE (Scale)	(a)	(b)
DIST	-114.0 (2.47)	-24.0 (0.39)
DISTSQ (10^{-6})	16.1 (6.62)	9.79 (3.41)
GEN	139.0 (3.23)	61.9 (1.01)
GENSQ (10^{-6})	11.8 (6.06)	8.40 (3.50)
DIST·GEN (10^{-6})	-27.6 (6.38)	-18.0 (3.38)
FCGEN (10^6)	-50.4 (1.04)	75.3 (1.53)
PURCH	-95.7 (2.84)	-24.9 (0.76)
PUB·DIST		170.4 (0.19)
PUB·GEN		-118.6 (0.13)
PUB·D·G (10^{-6})		18.8 (2.40)
PUB·PURCH		172.9 (0.21)
PUBLIC	-178.0 (2.87)	
COMP (10^6)	-50.1 (0.88)	346.0 (3.67)
COMP·DIST		-34.0 (4.79)
POOLMEM (10^6)	46.1 (0.95)	-13.5 (0.30)
HCSUB (10^6)	59.0 (0.76)	-98.4 (1.27)
INCENTR (10^6)	13.1 (0.19)	110.0 (1.88)
GASUTIL (10^6)	-16.3 (0.50)	46.1 (1.21)
NUCLEAR (10^6)	463.0 (2.82)	526.0 (2.72)

Table 6.1. COST EQUATION (Continued)		
VARIABLE (Scale)	**(b)**	**(b)**
HYDRO (10⁶)	57.7 (0.74)	51.8 (0.60)
OTHER (10⁶)	13.1 (0.18)	-30.2 (0.48)
HIVOLT (10⁶)	-18.4 (0.24)	112.0 (1.39)
RESSIZE (10⁶)	1.22 (0.21)	11.3 (1.98)
COMSIZE (10⁶)	0.12 (0.28)	0.58 (1.57)
INDSIZE	-73.3 (0.29)	-73.8 (0.34)
PRFSTE (10⁶)	2.35 (1.77)	0.68 (0.55)
PRFNUC (10⁶)	10.6 (2.67)	0.43 (0.05)
WAGE (10⁶)	8.60 (0.26)	-2.2 (0.08)
COSTCAP (10⁶)	397.0 (0.95)	233.0 (0.67)
REGIONS	*	*
CONSTANT	132.0 (0.64)	-235.0 (1.23)
R²	0.786	0.853
F	11.81	15.3
N	543	543

Notes: t-statistics in parentheses.
       * indicates dummy variables included.

convex, as before, with most coefficients statistically significant. The interaction term *DIST•GEN* that measures vertical economies between stages of production is negative and highly significant in both, confirming the existence of vertical cost complementarities. As before, vertical economies offset diseconomies of scale in generation and distribution, with the net effect of the quadratic and interaction terms again almost exactly offsetting. In all of these respects, these equations replicate that reported in Chapter 4.

With respect to the various structural and institutional features included, column (a) relies upon fixed-effects variables to capture the essence of their impact on costs. Public ownership (*PUBLIC*) is associated with significant cost reduction relative to IOUs, whereas the impact of competition, while negative, is not statistically significant. Pool membership, holding company subsidiaries, and incentive regulation are all found to be without significant effect. These results therefore confirm previous findings that costs are lower for publicly owned utilities. Other effects are statistically much weaker.

Certain changes emerge in the extension of this model in column (b). As expected, the proliferation of interactions with public ownership, some of which are now endogenous, results in several insignificant coefficient estimates. Only *PUB•D•G* is statistically significant. Its positive coefficient indicates that publicly owned utilities achieve much smaller vertical economies than do IOUs. Also as before, the interaction between output and competition (*DIST•COMP*) is negative and significant, whereas the fixed cost term associated with competition (*COMP*) is positive. Importantly, however, and while reduced in magnitude in column (b), the estimated coefficients on the output variables largely perform as in all earlier regressions.

The remaining variables in this regression either are very similar to those in the single-equation estimation or simply are of lesser interest. They will not be discussed further. Both equations in Table 6.1 have high pseudo-R^2 [3] and significant F-statistics, indicating substantial explanatory power. Overall, particularly given the complex specifications, these results are broadly consistent with earlier findings in the simpler OLS models.

Table 6.2 goes on to demand function estimation. The results are very similar to those in the markup model reported in Table 5.3. Quantity demanded is negatively influenced by *PRICE*, with a t-value of 4.02 and a coefficient now implying a demand elasticity of 0.62. Numbers of customers by segment, per capita income, natural gas prices by segment, and climate conditions all play nearly identical roles in influencing aggregate demand for electrical power. Demand behavior is well captured by this model.

The various influences on price are estimated in Table 6.3 in results that strongly resemble previous findings. Price is overwhelmingly determined by average cost, with a coefficient (.982) that now is statistically indistinguishable from the pass-through value of unity. Both tax rates and the proportion of high-voltage power also significantly affect price in the expected manner. As before, public ownership lowers final price, by an estimated 0.16 cents per kwh, or 2.7 percent, relative to IOU pricing. This effect is similar in magnitude to that previously found, but now holding constant the same numerous set of other influences and netting out any simultaneous effect of ownership on costs.

The fixed effects variable for duopoly utilities also confirms results found in earlier analysis. Duopoly utilities operate with significantly lower prices than those in monopoly settings, the difference a very substantial 0.46 cents per kwh,

Table 6.2. DEMAND EQUATION	
VARIABLE (Scale)	**(a)**
PRAVG (10^6)	-50.1 (4.02)
RESCUST	29.3 (17.3)
COMCUST	-56.9 (4.33)
INDCUST	467 (7.47)
INCOMPC	47.4 (0.66)
PRGASR	-0.23 (0.81)
PRGASC	0.80 (1.88)
PRGASI	-0.48 (1.85)
HEATDAY	160 (1.51)
COOLDAY	983 (2.79)
RESALE	1.57 (19.0)
CONSTANT (10^6)	-0.50 (0.30)
R^2	0.944
F	818
N	543

Note: t-statistics in parentheses.

or 7.8 percent. This, too, is quite similar to that estimated in the single-equation model. Neither public election of state utility commissioners nor incentive regulation appears to play a important role in pricing in this model, nor in the earlier one.

The feedbacks captured by this simultaneous-equations framework are conceptually important, as discussed at the outset. Yet, they do not substantially alter the estimated impacts of public ownership and competition previously found.

Table 6.3. PRICE EQUATION	
VARIABLE (Scale)	(a)
AVGCOST	0.982 (42.9)
TAXRATE	0.647 (10.4)
PCTHIVO	0.0052 (3.96)
PUBLIC	-0.0016 (1.97)
DUOPOLY	-0.0046 (2.78)
ELECTED	-0.0006 (0.47)
INCENTR	0.0002 (0.14)
CONSTANT	0.0056 (3.28)
R^2	0.865
F	402
N	543

Note: t-statistics in parentheses.

Both price and costs are found to be significantly lower under public ownership and in the presence of competition, with most effects statistically significant and their magnitudes plausible and of practical importance.

The sole previous study with any similarities to the present model is due to Hollas, Stansell, and Claggett (1992). They analyze costs and prices for municipally owned and *cooperative* distribution utilities, and also disaggregate demand into residential, commercial, and industrial customer classes. While the model has somequestionable features,[4] it is nonetheless worth noting that their study finds no significant cost effect from public ownership relative to cooperative distribution and significantly lower prices for residential and commercial users of power.

In summary, allowing for simultaneous determination of costs and prices and the effect of public ownership and competition on each strengthens the basis for the persistent findings of this study. Those findings are that, relative to regulated monopoly and private ownership, both public ownership and competitive distribution yield superior cost and price performanance.

6.2 Background to Customer Class Effects

Although most investigations into utility pricing focus on a single average price, a few studies have distinguished prices by customer classes. These permit more detailed examination of the purposes and effects of alternative pricing institutions on residential, commercial, and industrial users of electric power. Generally, these studies do find significant differences.

Meyer and Leland (1980) and Primeaux and Nelson (1980) both find margins[5] for residential power to be significantly larger than those for industrial power — or, in Meyer and Leland, for industrial plus commercial power. Primeaux and Nelson's calculation is based on marginal costs derived from estimated cost functions, whereas Meyer and Leland rely on direct measures of accounting costs. The latter also test whether customer class prices equal Ramsey or second-best prices.[6] They reject the latter possibility, reporting that low-elasticity residential customers were relatively favored by actual pricing in 1969, although the same was not true in 1974.

Primeaux and Nelson, by contrast, take *equal* margins for all customer classes as their benchmark, rather than efficiency-maximizing Ramsey prices. Not surprisingly, they find margins to be unequal, and conclude that residential customers must be facing adverse price discrimination. They also test whether price is yet more favorable to the industrial group where the latter is more important in the utility's customer base, but they find no support for this hypothesis.

In order to deduce marginal costs and ultimately margins by customer class, Hayashi, Sevier, and Trapani (1985) and Nelson, Roberts, and Tromp (1987) estimate translog cost functions. Both find residential margins to be higher than industrial margins, with commercial margins varying widely in between. Both studies reject Ramsey pricing, in particular, finding residential rates below the levels required and commercial rates above. Hayashi et al. report industrial rates too low, but somewhat more plausibly (given elasticities), Nelson et al. find them too high.

Only a few studies have examined the impact of public versus private ownership on electric utility rate structure. As with much else, the seminal work in this area is Peltzman's analysis (1971). Relying on political and voting theory, Peltzman argues that public enterprise pricing will favor the modal voting block, that pricing to the favored group will be more uniform (i.e., fewer within-group prices), and, as previously noted, that prices will be less closely related to underlying costs. His data for most of these issues are weak and contradict his hypotheses as often as support them. Yet his framework has proved fruitful for later work.

Utilizing translog cost function estimation for public and private utilities, a study by Hayashi, Sevier, and Trapani (1987) directly incorporates possible regulatory distortions in utilities' marginal costs. Among the principal findings is rejection of Ramsey pricing for both publicly owned utilities and IOUs. The latter have significantly higher margins for residential and commercial customers than for industrial users (though margins converge in the oil price shock year of 1980). As in the other Hayashi et al. study, residential customers of public power face *higher* margins than IOU residential customers, while industrial customers enjoy relatively more favorable treatment.

Despite some anomalies, this literature illustrates the fact that the three customer classes generally face different prices (or margins) under public ownership than under private ownership. The possibility that this pattern reflects deliberately different pricing practices, as well as simply controlling for variation in service mix, motivates this further inquiry into price effects.

6.3 Evidence by Customer Class and Mode of Ownership

The model required to examine pricing by customer class is a relatively straightforward generalization of that previously used. Instead of a single pricing equation, now one each for residential customers, commercial customers, and industrial customers is defined. With respect to demand, again three such functions are specified. A new identity sums the quantities sold in the three retail market segments plus sales for resale (which continues to be treated as exogenous) to total output. Costs are a function of total output, a function quite similar to that previously employed. The model therefore consists of seven behavioral equations — three demand functions, three pricing equations, plus total costs — together with identities for average cost[7] and for total quantity.

The results from estimating the total cost function are essentially the same as reported in Table 4.7 and will not be reproduced here. Rather, we turn our attention to the more interesting demand and pricing equations in Tables 6.4 and 6.5, respectively. As is evident in Table 6.4, there are fewer observations on industrial demand than for residential or commercial, with more than 40 utilities — all publicly owned — having no industrial users at all.[8] In addition to standard control variables, commercial demand is specified to be a function of the scope of the retail market served by commercial users of electric power (Hayashi et al., 1987). Hence, the number of residential customers is included in the commercial demand function. Consistent with other studies, climate conditions are omitted from the industrial demand function.

In all three equations, the quantities of residential, commercial, and

Table 6.4. DISAGGREGATED DEMAND			
VARIABLE (Scale)	**(a) Residential**	**(b) Commercial**	**(c) Industrial**
PR_ (10^6)	-2.56 (5.90)	-11.5 (3.68)	-25.8 (2.27)
RESCUST	8.56 (60.5)	4.41 (8.48)	
COMCUST		31.4 (7.74)	
INDCUST			907 (18.1)
INCOMPC	-12.4 (0.44)		
PRGASR (10^4)	17.7 (2.41)		
PRGASC (10^4)		9.14 (1.22)	
PRGASI (10^4)			41.0 (1.83)
HEATDAY	100. (2.32)	19.4 (0.61)	
COOLDAY	520.0 (4.36)	230 (2.72)	
CONSTANT (10^4)	-8.31 (0.17)	-6.27 (0.14)	62.8 (0.81)
R^2	0.885	0.938	0.400
F	690	1360	111
N	543	542	501

Note: t-statistics in parentheses.

industrial power demanded are negatively related to own price, positively related to the number of segment customers, positively related to the price of natural gas, and positively related to climate conditions, as would be expected. In addition, residential population as a measure of the retail market is positively related to demand for commercial power. Virtually all the estimated coefficients are statistically significant or very nearly so, and the overall explanatory power of these regressions is quite high. Only per capita income fails to emerge in the equation for residential demand, but it is statistically insignificant.

These results conform to theory in all respects and improve on those found in estimating an aggregate demand function. Here, for example, the roles for customer numbers and natural gas prices are much clarified. The coefficients on own price can be used to obtain implied demand elasticities. That for residential demand is .12, while commercial users' elasticity is .62, and that for industrial demand, .84. These estimates lie within the ranges for segment demand elasticities in the literature — perhaps a bit smaller for residential demand than what Bohi (1980, p. 60) terms "the consensus estimate...[of] around .2" and a bit larger in the case of commercial demand elasticity.[9]

This pattern, with residential demand the least elastic and industrial demand the most elastic (but still inelastic), undoubtedly reflects the differential access to and ability to utilize alternatives to utility-supplied power by each customer class. For industrial users, the alternatives may include self-generation, wheeling, and, in the longer term, relocation. Residential customers have fewer viable options.

Regressions for residential, commercial, and industrial pricing appear in the three columns of Table 6.5. In each case, the two cost variables, *AVGCOST* and *TAXRATE*, are positive and highly significant explanators of segment price. The coefficients on *AVGCOST* are statistically indistinguishable from one, indicating dollar-for-dollar pass-through of incurred average costs. The effect of incentive regulation is weak or perverse, presumably (as before) reflecting the fact that it is often instituted where price is unacceptably high. In these respects, the results corroborate findings from the earlier aggregate model estimation.

More varied and intriguing are the results on public ownership, duopoly competition, and popular election of utility commissioners. As shown in column (a), public ownership has a very large and significant effect on the price of residential power relative to IOU prices. The differential, 1.20 cents per kwh, is fully 15.4 percent less than the average residential price from investor-owned utilities. This represents a huge effect of public ownership for residential customers, suggesting perhaps that much of the ownership effect previously detected for average price may be concentrated in this segment.

Confirming this, column (b) reveals a price reduction of 0.34 per kwh for commercial users under public ownership, considerably smaller than the advantage for residential users. This effect is nonetheless significant and constitutes a 4.8 percent differential relative to IOU pricing. Clearly, both of these customer classes realize significant price benefits from public ownership.

A rather different story holds for industrial users. Their power actually appears to be priced slightly *higher* under public ownership, though not quite significantly so (t = 1.29). The estimated differential of 0.16 per kwh is 3.1 percent of industrial price from IOUs. In fact, a large and significant price differential to

Table 6.5. DISAGGREGATED PRICES			
VARIABLE (Scale)	**(a) Residential**	**(b) Commercial**	**(c) Industrial**
AVGCOST	0.944 (27.8)	1.07 (26.0)	0.952 (27.1)
TAXRATE	0.744 (8.12)	0.384 (3.46)	0.306 (3.16)
PUBLIC	-0.0120 (9.88)	-0.0034 (2.29)	0.0016 (1.29)
DUOPOLY	-0.0034 (1.38)	-0.0059 (1.97)	-0.0047 (1.90)
ELECTED	-0.0028 (1.51)	0.0011 (0.48)	-0.0020 (1.06)
INCENTR	0.0027 (1.55)	-0.0022 (1.02)	-0.0011 (0.63)
CONSTANT	0.020 (9.63)	0.009 (3.38)	0.0030 (1.38)
R^2	0.754	0.649	0.680
F	234	136	145
N	543	542	501

Note: t-statistics in parentheses.

industrial users would be unlikely, given their various alternatives to local-utility power noted above. Perhaps the better interpretation of this result is that industrial power price simply does not differ much by ownership mode.

These estimated effects of ownership are quite striking. They suggest that ownership does indeed result in a price advantage, but that the overall average combines a huge benefit for residential users with a more modest effect for commercial demand, and little or none for industrial demand. Indeed, the latter appear possibly even to be the source of some revenue transfer among market segments, from industrial to residential.

This same regression in Table 6.5 also finds lower prices in the presence of duopoly competition. The amount ranges from 0.34 to 0.59 cents per kwh, or 5.7 percent for the residential class, 9.4 percent for commercial users, and 8.1 percent in the industrial segment. These effects, too, are quite large, and both of the latter two estimates have t-values approaching statistical significance. These results suggest a fairly strong and systematic effect of duopoly competition throughout all market segments. Competition in retail distribution would appear to matter in segment-specific pricing at least as much as it does in costs.

Popular election of state utility commissioners is not quite significantly related to pricing in any of these segment regressions, but the pattern of estimated coefficients on the *ELECTED* variable is nonetheless noteworthy. The coefficient is largest and closest to being significant in the case of residential customers. Its value there implies a 0.28 cents per kwh or 3.6 percent price reduction for investor-owned utilities[10] in states that elect their commissioners. The estimated effect for industrial customers is of roughly the same magnitude (0.20 cents, or 3.9 percent of average IOU price to the segment), but it is statistically weaker. That for commercial customers is insignificant and even positively signed.

Both public ownership and popular election of utility commissioners, then, result in significantly lower prices for residential customers of electric power. Interestingly, both ownership and election are institutional choices influenced by pressure from the residential constituency. One might even suppose that the constituency chooses between these alternatives based on whichever is easier to achieve and more likely to yield the desired price objective. This interpretation will be taken up in greater detail in the next chapter.

6.4 Summary

Two major extensions of the previous analysis have occupied this chapter. This simultaneous-equations model is intended to explain the joint determination of price, quantity, and costs of electric utilities. The essential mechanism is one in which costs are determined by the utility based on its output, input prices, and other cost shifters. Output equals consumer demand which in turn is influenced by price. Finally, as a practical matter, price is based on average costs. This approach represents a substantial generalization of the single-equation framework. All important variables are now jointly and simultaneously determined, as they undoubtedly are in practice. This allows the price and cost impact of each variable of interest to be isolated and quantified. The results reflect the interdependency of these variables, and at the same time confirm the basic results on ownership, competition, and vertical integration previously obtained.

The specification of separate demand and pricing equations for residential, commercial, and industrial customer classes also confirms the essence of previous findings, but it has also uncovered some novel features. Those include the much greater effect of public ownership on residential prices, as compared to commercial or industrial prices; the clear effect of competition on all segments; and the suggestion that popular election of utility commissioners has an effect, especially on residential prices.

Notes

1. Among alternative techniques for estimating simultaneous equations, the principle reason for choosing instrumental variables is the need to estimate a disaggregated model, as will be discussed in the next section. An immediate advantage of instrumental variables is that it resolves a problem with overall average price of electric power. Since average price is often derived from a declining schedule of prices in conjunction with demand conditions, price is to some extent endogenous. Instrumenting for price resolves this problem. See Taylor (1975), Halvorsen (1975), and Hayashi et al. (1987).

2. These were previously denoted *DIST•POOL, DIST•HCSUB,* and *DIST•INCENT.* Each of these interactions would be an additional endogenous variable and highly collinear with distribution output. Although their omission should not greatly affect other variables, we avoid reliance upon this regression for insight into the cost effects of those variables. Those effects have been extensively examined earlier.

3. Two-stage least squares does not produce the R^2 conventional in OLS. Pseudo-R^2 is an analogous construct that measures the percent of total variation explained by right-hand side variables.

4. For example, a variable for municipal versus cooperative ownership is included in the *demand* equation. The results also imply implausibly high elasticities. And the effort to be faithful to the translog form results in large numbers of endogenous interaction terms in the cost function, leading to dozens of insignificant coefficients.

5. Margins are often used instead of prices themselves in order to allow for differences in class-specific costs: The costs of supplying power to residential users is larger, for example, due to lower required voltage, greater line losses, and larger service and billing expenses.

6. Ramsey prices exceed marginal costs in inverse proportion to demand elasticity in each market segment. This minimizes the efficiency loss, while achieving breakeven operation. See, for example, Berg and Tschirhart (1988) for further discussion.

7. It should be noted that marginal cost estimates have not been developed here, but only average costs as the basis for pricing. This still allows comparisons of class-specific prices between public and private utilities, even if the price-cost differential may not be precisely estimated. Also, since segment demands are specified explicitly, the variable measuring the percent of high-voltage output HIVOLT is now treated as endogenous in the cost function and is omitted altogether from the disaggregated pricing equations.

8. These different numbers of observations by segment ultimately dictate use of two-stage, least-squares estimation. Systems methods would reduce the total number of observations to the smallest number in any equation (i.e., 501).

9. In addition to the numerous studies summarized in the surveys by Taylor and by Bohi, see Berndt and Wood (1975), Halvorsen (1975), Beierlein et al. (1981), and Hayashi et al. Bohi's conclusion, finding "wide disparities in the estimates reported", remains apt.

10. Recall that publicly owned utilities are generally not regulated by state commissions.

Chapter 7

THE POLITICAL ECONOMY OF OWNERSHIP AND REGULATION

The evidence thus far has shown that public ownership confers both cost and price benefits, the latter primarily accruing to residential customers. Similarly, popular election of state utility commissioners is also associated with price reductions to residential users. While in each case either commercial or industrial customers stand to gain a bit from public ownership or popular election, it seems clear that the major beneficiaries of these alternatives to conventional regulation with appointed commissions are residential customers.

It may therefore be expected that residential consumers of electric power prefer either public ownership or popular election of commissioners, actively seek these alternatives, and succeed in their efforts under sufficiently favorable circumstances. This chapter conducts an exploratory examination into these choices. Specifically, it examines the informational, organizational, and countervailing forces giving rise to public versus private ownership of electric utilities and to popular election of public utility commissioners.

Two approaches to these questions are pursued. The first draws on economic theories of regulation and of ownership for insight into the determinants of public versus private ownership and of popular election of state utility commissioners. Characteristics of jurisdictions favoring either regime are identified and tested to see whether they in fact have motivated adoption. As part of this effort, the determinants of contributions and service flows between publicly owned utilities and their municipalities are also examined. The second approach focuses on those jurisdictions that have switched ownership modes for direct evidence on the factors motivating the choice of public versus private ownership.

7.1 Theories of Public versus Private Ownership

Virtually all policy actions differentially benefit various constituencies. In order to identify favored constituencies a priori and to describe the mechanism by which those constituencies secure their chosen outcome, we begin with Stigler's (1971) pathbreaking work on economic regulation. While Stigler did not discuss public ownership as an alternative strategy, his insight is nonetheless a proposal: Individuals with lower transactions costs of organizing into interest groups and with a greater stake in the outcome will be more successful in securing their objective through nonmarket processes. Stigler proposed several observable characteristics contributing to such success: high per capita benefits, smaller and more cohesive groups, etc.

Subsequent research has significantly advanced this theory. Peltzman (1976) in fact discusses public ownership and regulation as alternatives. He begins by assuming that any politically determined price short of maximum profitability requires a tax increase to offset lost profits. That implies a trade-off between the political benefits of lower prices and the political costs of higher taxes, both measured in terms of votes gained and lost by the decision maker. The relative effects depend upon the income elasticity of the product versus its tax elasticity. Becker (1983) sets out "influence functions" of competing interest groups and identifies a "political equilibrium" between groups as the probable outcome. An extensive literature on political and regulatory processes has arisen from these theories, substantially confirming their major implications.[1]

Theory going back to Alchian and Demsetz (1972) discusses the nature of ownership and management and the circumstances that motivate various relationships between the two. As noted earlier, Hansmann (1988) has a particularly useful review. He argues that cooperatives — his example happens to be rural electrical coops — are devised by consumers to avoid the costs of monopoly and regulation. Hansmann further contends that coops can be expected to arise in circumstances of greater population stability and homogeneity, such as rural areas. Nonprofits generally are said to be responses to an extreme asymmetry of information and interests between the enterprise and a single class of customers. If that class (residential customers in the present context) is not given effective ownership of the firm, Hansmann argues, they would be subject to exploitation.

Among other recent studies, Hart and Moore (1994) examine member cooperative versus outside ownership as alternative governance mechanisms.[2] They take a cooperative to be an institution with assets controlled by the democratic vote of its members, whereas outside ownership simply means profit-maximizing behavior. Hart and Moore establish that greater skew[3] to member preferences implies that profit-maximization is more efficient, since it is more responsive to "marginal" members, who, in turn, determine optimality conditions. By contrast, the cooperative form is responsive to the *median* member voter. As the latter's interests diverge from those of the mean member, total surplus will be sacrificed by the cooperative's actions. Such nonprofit institutions may therefore be expected to arise in response to preferences that are relatively less skewed.

Bernard and Roland (1995) adapt Peltzman's framework regarding the distribution of costs and benefits from public ownership to Canadian electric utilities. They observe hydropower utilities setting price well below marginal (fossil-fuel) generation costs. Assuming that the revenue shortfall is made up from general tax revenues derived by a proportional income tax, Bernard and Roland derive the elasticity conditions under which the median voter prefers below-cost service subsidized from the majority of (higher income) taxpayers.[4] They offer some suggestive evidence from the experience of Hydro-Quebec and other provincial electric utilities.

A few studies have now been concerned with the fact that these alternative regimes may themselves be the result of choices. In particular, if the choice of public versus private ownership, election versus appointment of utility commissioners, incentive versus rate-of-return regulation and so forth is motivated by differential past performance or by distinctive potential for future performance benefits, any ex post statistical association between these institutions and performance may not reflect causality. Put differently, if public ownership is adopted in a select subset of all jurisdictions, it may be the underlying characteristics of those jurisdictions rather than public ownership that results in measured superior performance.

Recent studies in this vein include Berg and Jeong (1991) and St. Marie (1996).[5] Each examines incentive regulation for electric utilities, explicitly noting and testing for the performance-based motivation for adopting incentive regulation. Neither finds that incentive regulation significantly alters performance, whether or not adoption is made endogenous. Although unrelated to electric power, work by S. Schmidt (1995) carries the modeling of institutional choice the farthest. Schmidt notes that while most studies show greater efficiency for publicly owned mass transit, none really corrects for the fact that public and private systems arise in different types of localities. After correcting for this selection bias (specifically, jurisdictions favoring lower rider fares are more likely to adopt public ownership and then to subsidize fares from local tax revenues), Schmidt finds little difference between public and private transit systems in service or cost.[6]

Although resolution of questions about causality is beyond the scope of this study, available data do cast some light on observable determinants of public ownership, popular election of utility commissioners, and subsidy flows between publicly owned utilities and their municipalities. The next section examines some of these determinants of the choice of ownership and related institutional regimes.

7.2 The Determinants of Ownership and Election

The economic theory of regulation, voting theory, and previous analyses suggest various determinants of public versus private ownership of electric utilities and of popular election versus appointment of utility commissioners. These fall into two categories — namely, the potential gain to some constituency that is thereby motivated to pursue the favorable regime, and organization costs that affect that constituency's ability actually to secure such benefits. We take these up in order.

With respect to potential gains, it is clear that residential customers should represent the major interest group favoring public ownership. We would therefore expect those jurisdictions where they are relatively more numerous to have a higher frequency of public ownership. Residential customers' stake in lower electric prices also depends upon some measure of their consumption of or expenditure on electric power. Past results also suggest that commercial customers may benefit from public ownership, but to a much smaller degree, while industrial users neither gain nor lose much and should therefore be largely indifferent.

The actual variables representing these and other factors in the analysis are largely familiar from previous work. *PCTRES* is simply residential customers as a percentage of total customers. Average electricity consumption for residential, commercial, and industrial customers is specified as each segment's consumption divided by the number of segment customers and is denoted (as before) by *RESSIZE, COMSIZE*, and *INDSIZE*, respectively.

The second category of influences on ownership choice concerns the actual ability of the potentially benefiting group to organize in pursuit of their objectives. Theory implies that large groups and jurisdictions entail higher organization costs and thereby reduce the likelihood of success. By contrast, higher density of customers would reduce the costs of organizing that constituency and thus would favor public ownership. And lastly, per capita income may play a role in organization, although the direction of its effect is ambiguous: A higher income population might be more inclined to mobilize politically, but alternatively, its higher opportunity cost may outweigh that inclination.

In the empirical specification, utility size is captured simply by the total of residential, commercial, and industrial users, denoted *CUSTOM*. *INCOMPC* represents average income per capita and, while the data are not available for all utilities, *DENSITY* measures the number of residential customers per distribution mile.[7] With respect to heterogeneity or skewness of customer characteristics, no data are available. It should also be noted that fixed-effects variables for regions are included to capture any other unmeasured influences on ownership that might differ by region of the country, such as attitudes towards public ownership or ease of formation of publicly owned utilities.

The dependent variable in the model is the zero-one variable *PUBLIC*, requiring probit analysis instead of ordinary regression techniques. The results are reported in Table 7.1. The most important explanator of public ownership would appear to be the total number of customers in the service territory. The strongly significant negative sign on *CUSTOM* supports the proposition that smaller jurisdictions find it easier to organize and secure public ownership.[8] Somewhat unexpectedly, *PCTRES* fails to emerge in this or other specifications as significantly related to the choice of public ownership. This might simply be due to the previously observed fact that the fraction of customers that are residential does not appear to differ much between public and private utilities. It does *not* imply that public ownership is unrelated to the numbers of residential customers. Their numbers are highly correlated with *CUSTOM*, simply making it difficult to disentangle their separate roles.

The remaining variables in this specification largely conform to expectations. Average residential electricity usage is positively related to public ownership, indicating that areas where residential customers purchase more electricity are indeed more likely to adopt low-cost public power. Greater commercial customer usage is also, but more weakly, related, while larger industrial usage disfavors public ownership. Each of these effects closely conforms to the previous findings with respect to the impact of public ownership on the rates charged to the three segments. Lastly, per capita income has a positive and significant relationship to *PUBLIC*. This suggests that, on balance, it is more affluent areas that tend to have publicly owned utilities.

The model estimated in column (b) differs from that in column (a) by including the variable for the customer density, but at the cost of a reduction in the number of observations. *DENSITY* is intended to capture the organizational advantages of customer proximity that might benefit movements for public ownership, although it may also reflect the degree of customer homogeneity. This variable proves to be statistically quite important, with a significantly higher probability of public ownership in denser areas. Most, but not all, of the other variables retain their signs and significance.

In summary, these probit analyses demonstrate clearly the importance of the size of jurisdiction (*CUSTOM*) in the choice of public versus private ownership, as well as the roles for per capita income, customer density, and, somewhat less reliably, average residential electricity consumption. Most of these factors are readily interpreted in terms of constituencies' choices of governance mechanisms in general and public ownership in particular. A high percentage of overall variation is explained, with pseudo-R^2s in excess of .50.

Table 7.1. DETERMINANTS OF PUBLIC OWNERSHIP AND ELECTION OF UTILITY COMMISSIONERS				
VARIABLE (scale)	(a)	(b)	(c)	(d)
CUSTOM (10⁻⁶)	6.03 (10.7)	6.17 (10.1)	-.011 (.26)	-.032 (.75)
PCTRES	.200 (.07)	-.519 (.16)	-.556 (.40)	.801 (.51)
RESSIZE	.053 (1.52)	.052 (1.36)	.023 (1.61)	.032 (2.04)
COMSIZE (10⁻⁴)	5.63 (.20)	-12.9 (.54)	10.2 (.76)	10.8 (.77)
INDSIZE (10⁻⁶)	-1.63 (1.22)	2.54 (.39)	-.117 (.35)	-6.48 (1.54)
INCOMPC (10⁻⁴)	1.94 (4.00)	1.72 (3.31)	-.729 (5.10)	-.675 (4.56)
DENSITY		.0061 (3.66)		.0005 (1.24)
REGIONS	*	*	*	*
CONSTANT	-2.66 (1.04)	-2.02 (.70)	1.56 (1.28)	.196 (.14)
N	543	507	147	147
R²	.547	.563	.490	.491
Chi Sq/F	347	325	9.06	7.34

Notes: t-statistics in parentheses.
* Indicates dummy variables included.

Next we turn to the issue of popular election versus appointment of state public utility commissioners. Unlike public ownership, popular election covers all utilities in a *state* so that certain utility-specific characteristics — for example, size or density — can be expected to be less important. In addition, earlier evidence suggested that the major benefits of popular election accrue to residential customers, with some to industrial users as well. The most pronounced effect on popular election would thus be expected to result in cases of greater electricity usage by residential customers. As before, the role of per capita income is ambiguous.

Many of the variables determining popular election of state utility commissioners may therefore not differ greatly from those for public ownership. Columns (c) and (d) of Table 7.1 report the results of an exploratory analysis of these determinants. Only investor-owned utilities are included, since it is their customers whose rates are overseen by state utility commissions. As before, the second of these two regressions includes *DENSITY* and is therefore based on somewhat fewer observations.[9]

Some interesting differences emerge between the determinants of popular election and public ownership. Much as predicted, utility size — measured by number of customers — does not affect the likelihood of popular election, since the latter is a statewide issue. For similar reasons, the effect of *DENSITY* in column (d) is much smaller and statistically insignificant, as compared to its impact on public ownership. *PCTRES* plays no measurable role in either decision.

With respect to the other variables, average consumption by residential customers *RESSIZE* continues to be the only significant (or nearly significant) usage factor increasing the likelihood of popular election. This is consistent with the previous finding that they benefit substantially from such an arrangement. Average industrial usage is negatively related to election, however, despite evidence that industrial rates are reduced as well. This may reflect misgivings by industrial customers as to the consequences of popular election. Strikingly, per capita income now has a negative and significant coefficient, implying that popular election arises in relatively less affluent jurisdictions. This is in sharp contrast to public ownership, suggesting that these strategies for securing rate benefits arise under precisely the opposite income conditions.

Additional insight into the public/private choice may be obtained by examining the proposition that public ownership is used to redistribute benefits between the utility and its town, rather than the utility and its customers. To the extent that transfers of funds or services are important, public ownership might be motivated by some group's interest in such transfers. Although earlier data analysis showed that contributions and services on average account for a small percentage of utilities' total costs or revenues,[10] they may still represent a significant flow to some jurisdictions.

Table 7.2 relates the percentage of total costs represented by contributions and services to a series of explanatory variables. Column (a) examines contributions and services *by* the utility to the town, (b) considers such contributions and services by the town *to* the utility, and (c) represents *net* contributions and services, that is, column (a) less column (b). Observations are limited to the 396 publicly owned utilities, all of which make or receive contributions or services.

The most important explanator of contributions and services (CS) by a utility appears to be average residential usage. Interestingly, it is negatively related, implying that higher usage is associated with smaller contributions by the utility. In conjunction with previous results, this suggests that intensive users prefer to realize their benefit of public ownership in the form of rate reduction, which is necessarily reduced to the extent that *CS* by the utility increases. Also noteworthy is the fact that both commercial usage and industrial usage are positively related to greater *CS*, although the effect for the industrial segment is insignificant. It may be that while residential customers prefer (or can secure) rate benefits, other customer classes find non-rate benefits (*CS*) easier to obtain.

The other variables are *PCTRES* and *INCOMPC*.[11] The fact that the percentage of all customers represented by residential customers is insignificant in this regression may simply reflect the fact that contributions and services benefit all groups to some extent. On the other hand, per capita income is positively related to contributions and services, implying that (as with public ownership itself) higher income jurisdictions strive to extract greater nonrate benefits from public systems. This is consistent with the view that *CS* flows reduce local taxes to the differential advantage of higher income residents.

The results in column (b) of fitting this model to contributions and services *from* the town *to* the utility have much less statistical significance and explanatory power. The only noteworthy feature is that, while statistically insignificant, greater usages by all three customer classes disfavor flows from the town.

The final column demonstrates that these variables explain much of the difference between percentage contributions from and to the utility, that is, the net flow. In this, they track the column (a) regression on *CS* itself. As before, average residential usage is strongly associated with lower net flows, while commercial and industrial usage rates are (less strongly) associated with larger net *CS*. Also as before, higher per capita income increases net *CS* and the percentage of residential customers makes no difference.

Taken together, these results suggest that there are important causal determinants of public ownership, popular election of utility commissioners, and contributions and services. It is clear that public ownership arises in circumstances characterized by certain size, income, and residential stake in power costs. Popular election appears also to arise when the stakes are substantial, but in the opposite income circumstances. With respect to contributions and services, the evidence suggests that higher income areas prefer *CS* by the utility, although greater residential usage does not. The latter probably reflects a preference for rate reductions instead.

Table 7.2. DETERMINANTS OF FLOWS OF CONTRIBUTIONS AND SERVICES			
Variable (scale)	(a) CS by Utility	(b) CS to Utility	(c) Net CS by Utility
PCTRES	.0014 (.01)	.0057 (.48)	-.0043 (.04)
RESSIZE (10^{-4})	-61.1 (5.10)	-.834 (.63)	-60.3 (5.11)
COMSIZE (10^{-4})	1.72 (1.95)	-.060 (.61)	1.78 (2.05)
INDSIZE (10^{-6})	358. (.44)	-.0057 (.64)	415. (.52)
INCOMPC (10^{-6})	4.43 (1.97)	18.6 (.75)	4.24 (1.92)
REGIONS	*	*	*
CONSTANT	.0036 (.04)	-.0061 (.56)	.0097 (.10)
N	396	396	396
R²	.121	.044	.127
F	4.05	1.35	4.27

Notes: t-statistics in parentheses.
  * Indicates dummy variables included.

7.2 Conversions to and from Public Ownership

The causes and effects of ownership mode may also be illuminated by examining those utilities that have converted from public to private status or vice versa. Tables 7.3 and 7.4 report compilations by the American Public Power Association of formations and of sellouts, respectively, of publicly owned utilities between 1980 and 1994. A total of 33 new publicly owned utilities were formed during this period, of which 17 were conversions from investor-owned to public status, the remainder arising primarily from some federal power supply arrangement.

Table 7.3. PUBLICLY OWNED UTILITES ESTABLISHED 1980 - 1994

Utility Name	State	Number of Meters
1981		
Chignik Electric	Alaska	60
Massena Electric Department	New York	8,390
1982		
City of Kotlik Utility	Alaska	122
Akutan Electric Utility	Alaska	39
Tuolumne County Public Power Agency	California	215
Trinity County Public Utility District	California	2,189
1983		
City of Thorne Bay Utilities	Alaska	306
St. Paul Municipal Electric Utility	Alaska	186
Kwig Power Co.	Alaska	85
City of Needles	California	2,500
Emerald People's Utility District	Oregon	13,657
1984		
City of Larsen Bay	Alaska	62
Columbia River People's Utility District	Oregon	7,310
1985		
City of Galena	Alaska	350
Page Electric Utility	Arizona	2,991
1986		
Town of Pickstown	South Dakota	63
San Marcos Electric Utility District	Texas	12,420
Strawberry Electric Service District	Utah	1,993
1987		
Town of Fredonia	Arizona	526
Electric District #8 of Maricopa County	Arizona	188
Reedy Creek Improvement District	Florida	658
Troy Power and Light	Montana	879
Kerrville Public Utility Board	Texas	15,845
Kanab City Corporation	Utah	1,200

Table 7.3. PUBLICLY OWNED UTILITES ESTABLISHED 1980 - 1994 (Continued)

Utility Name	State	Number of Meters
1988		
Hayfork Valley Public Utility District	California	724
Lassen Municipal Utility District	California	10,000
City of Scribner	Nebraska	549
San Saba Electric Utiltiy	Texas	1,713
City of Santa Clara	Utah	600
Washington City Electric System	Utah	1,600
1989		
Cldye Light and Power	Ohio	1,921
1990		
Byng Public Works Authority	Oklahoma	NA
1992		
Riverdale	North Dakota	150

Source: American Public Power Association, undated.

Seventeen total conversions represent about one conversion per year, a glacial pace of change in an industry of some 3,000 producers.

Newly formed public utilities tend to be quite small. They range in size from the 39 meters[12] of Akutan (Alaska) Electric to 15,800 for the Kerryville (Texas) Public Utility Board. Their average of about 2,800 meters represents less than 10 percent of the average number of customers for all publicly owned utilities in the present data set, and a yet smaller fraction of that for all utilities.

It is also noteworthy that 13 of the 17 former IOUs that converted to public ownership are offshoots of only five private utilities — Arizona Public Service, CP National, Pacific Gas & Electric, Pacific Power & Light, and Utah Power & Light. In addition, these 13 have similar years of formation. These facts suggest some common opportunity or motivation for conversion to publicly owned status.

The overall timing of these conversions deserves comment. Passage of the 1992 Energy Policy Act prompted speculation that the new transmission access mandates would cause communities to adopt municipalization and then shop for

Table 7.4. PUBLIC POWER SELLOUTS 1980 - 1994

Utility Name	State	Number of Meters
1980		
Jamestown	Kansas	244
Franklin	Lousiana	3,026
Melville	Louisiana	600
Tucumcari	New Mexico	3,000
Monticello	Utah	475
Upton	Wyoming	350
1981		
Lake Helen	Florida	715
Oswego	Kansas	1,001
Wilson	Kansas	500
Opelousas	Louisiana	5,890
Centerville	Maryland	NA
Valley Springs	South Dakota	Approx. 300
Canadian	Texas	1,053
Winters	Texas	875
1982		
Eldon	Iowa	625
Willis	Kansas	48
Princeville	North Carolina	650
1983		
Dubois	Idaho	205
Kelley	Iowa	NA
Ellis	Kansas	962
Goff	Kansas	962
Protection	Kansas	418
Tipton	Kansas	NA
Clayton	New Mexico	1,325

(Continued)

Utility Name	State	Number of Meters
1984		
Roland	Iowa	500
Crosbyton	Texas	913
Vernon	Texas	2,337
1985		
McKinley	Minnesota	60
Sargeant	Minnesota	52
Centreville	Mississippi	NA
Amorita	Oklahoma	80
Byron	Oklahoma	30
Carmen	Oklahoma	230
Cherokee	Oklahoma	1,441
Waleetka	Oklahoma	NA
Sonora	Texas	882
Deerfield	Wisconsin	486
1986		
Worley	Idaho	104
Seymour	Iowa	NA
Netawaka	Kansas	NA
Lewiston	Maine	NA
Iron Gate	Virginia	NA
1987		
Walsenburg	Colorado	2,100
Footville	Wisconsin	352
1988		
Coats	Kansas	75
Manchester	Oklahoma	105
1989		
Chelsa	Oklahoma	945

132

Utility Name	State	Number of Meters
1990		
Mahaska	Kansas	76
Lubec	Maine	1,300
1991		
Clarksville	Missouri	400
Pleasant Hill	Missouri	NA
1992		
Coffman Cove	Alaska	71
Eagle Village	Alaska	15
1993		
Sebring	Florida	12,278
DeSoto	Kansas	910
Year Unknown		
Paxton	Nebraska	305

Source: American Public Power Association, undated.

lower cost supply sources (Coopers & Lybrand, 1993). A significant countervailing force has been the 1987 "Rostenkowski Amendment" that restricted use of tax-exempt bonds for the purpose of financing takeovers of IOU facilities. In addition, there are a number of longstanding obstacles to the municipalization process (discussed further below). The net effect of these forces appears clear: Of the 33 total conversions to publicly owned status, only two have taken place during the 1990s, and none since 1992. Rather than dominating, the incentives and ability to undertake municipalization appear to have diminished in recent years.

During this same 1980 – 1994 period, as shown in Table 7.4, a total of 56 publicly owned utilities disappeared. Most of these represented privatizations, that is, the purchase of facilities by investor-owned utilities — typically large contiguous companies, sometimes even surrounding the public system. About 15 "disappearances" of publicly owned utilities entailed sellouts to cooperatives. These numbers imply an average rate of disappearance of about three per year to private status, plus another one to cooperative membership. Eight sellouts have occurred since 1990, considerably more than conversions in the other direction. Two of these postdated the 1992 Act, including the largest privatization, namely, that in Sebring, Florida. These 56 cases are geographically scattered quite widely, from Maine to New Mexico and Alaska. Kansas accounts for 12 cases and Oklahoma for an additional seven privatizations.

The publicly owned utilities that have disappeared are even smaller than new formations. These 56 average only about 1,000 meters, slightly more than one-third the size of new public systems. The net effect of these relative sizes and numbers of conversions implies that a total of about 56,000 utility customers (or meters) have abandoned public supply in this 15-year period, while 92,000 have converted to public ownership. Out of a national electric customer base of nearly 100 million, total conversions in *both* directions in the *entire* period have involved less than 0.2 percent of all customers.

This seemingly trivial incidence of conversion would seem to believe the attention given to the public-versus-private ownership issue. In reality, however, at any point in time, many more privatizations and municipalizations may be under active consideration. One source reported that in the single year 1989, "more than fifty localities contracted for feasibility studies to consider taking over private utility facilities or establishing local electric systems,"[13] and, more recently, 25 communities are said to be "actively pursuing conversion" (Salpukas, 1995). Obviously, most of these do not result in actual conversions, but this activity is a constant reminder that current institutional arrangements need not be permanent.

There are two reasons why the number of actual conversions to public ownership falls well short of the number under consideration: Some encounter insurmountable practical problems, while others "succeed" without actually undergoing the conversion process. The difficulties associated with conversion of an IOU to public ownership are many. Apart from the transmission access issues discussed earlier, the issues to be resolved include the legal authority to acquire an IOU's retail distribution facilities, the choice of method (nonrenewal or revocation of franchise, eminent domain, etc.), valuation of utility property, treatment of "stranded costs," and securing of financing.[14] As previously noted, financing has recently been complicated by restrictions on the use of tax-exempt bond financing of takeovers.

Some of these impediments may be at least partially circumvented by alternatives to complete takeover and replacement of a privately owned utility. Among these alternative strategies for municipalization are the following:

■Annexation of adjacent areas previously served by an IOU. Although this raises many of the same issues as do takeovers, the fact that the publicly owned system is already in existence simplifies some aspects of the transaction.

■Phased municipalization, wherein the new utility targets customers or areas for its initial efforts at conversion. This strategy has advantages in terms of phasing in costs, while initially securing greater revenue benefits. Sometimes, specific customers are not so much targets as the driving forces for municipalization. Industrial customers in particular may encourage their local municipalities to get into the electricity distribution business in order to shop for and secure lower cost power for themselves and, incidentally, for other customers.

■Construction of duplicative distribution facilities. While expensive, this strategy represents a still-permitted use of local bond financing after the 1987 legislation. Despite considerable speculation that more municipalities would resort to its use, this has not proved to be the case. There are, nonetheless, some longstanding examples of duplicative systems, previously discussed in the context of competition. A notable case is that of Cleveland Public Power, which has pursued a long-term strategy of gradually adding load and extending its distribution facilities into territory supplied by the private Cleveland Electric Illuminating.

A more recent example of duplicate facilities is provided by Clyde, Ohio. Beginning in 1986 and supported by its major local customer, Whirlpool, Clyde sought to purchase Toledo Edison's distribution system within the town and then to secure a lower cost power supply. Rebuffed, the town proceeded to build duplicative facilities (giving rise to the term, "Pole City"). It is offering service at a reported 25 percent savings, and most recently is seeking to force Toledo Edison to withdraw altogether (Salpukas).

■Minimal municipalization, or "municipalization lite." The 1992 Energy Policy Act made eligible for mandated transmission access any utility that owned distribution facilities. In the absence of a definition of the latter, Falls Church, Virginia, has proposed to install city-owned meters only, shop the aggregated load in the wholesale generation market, and then request transmission/distribution service from the surrounding IOU (Virginia Power) all the way to the meter. Whether this qualifies as a distribution utility under the law — and if not, what does — will ultimately be decided by the courts.[15]

If ownership of meters proves to satisfy the legal requirement, it might be expected that a greater number and a greater variety of minimal municipalizations would occur. Jurisdictions could then solicit bids from generators without the need for any substantial facilities investment, vastly simplifying the process now required.

The second reason why few threats or considerations of municipalization come to fruition is that the IOU frequently responds by offering rate concessions, sometimes to entire towns, sometimes to selected customers. This response results in the very rate benefits that public ownership was intended to achieve, but without the transition costs, and thus is often irresistible. For example, in 1989, the city of Marion, Ohio, was considering a proposal for takeover of Ohio Edison facilities, a proposal strongly backed by Marion Steel, which represented 40 percent of the city's load. When Ohio Edison offered the company an immediate 10 percent rate reduction and disclosed plans to make certain investments in the city, Marion Steel withdrew its support, and the Marion city council voted the proposal down.[16]

A similar episode transpired in Rome, Michigan, where the Ford Motor is the major customer. In April 1994, Ford sought to build support for municipalization by funding a preliminary analysis which predicted substantial savings for both residential and other customers. Ford was simultaneously negotiating directly with Detroit Edison for lower rates, however, and upon securing them in February 1995, withdrew its support for municipalization. That proposal was thereupon tabled.[17]

Given these experiences, Toledo Edison's effort to head off consideration of municipalization in Toledo, Ohio met with a somewhat unexpected reaction. After the city council voted to initiate a feasibility study in 1990, Toledo Edison offered to reduce rates by four percent if the study was dropped. This conditional offer was vehemently criticized as a "bribe," prompting rejection by the city. Most recently, Toledo's attention has been focused on phasing in a duplicative distribution system along the lines of Cleveland Public Power.

Anecdotal evidence regarding conversions *from* publicly owned status *to* private ownership suggest quite a different pattern. A number of these cases represent management failures under public ownership, leading to takeover by a nearby investor-owned system. Not infrequently, the finances or facilities of the public system had declined to the point that such a takeover has been welcomed, even sought, by local residents.

Typical experiences include those of Chelsea, Oklahoma, and Sebring, Florida.[18] In the former case, moneys were extracted from the utility for town use. From about $35 – 50,000 per year in the later 1970s, the annual transfer to the general fund totaled $227,000 in 1986. The utility's employees were said to spend more time on nonelectric matters for the town than on the utility and, not surprisingly, facilities deteriorated. By 1988, the town voted to sell their system to Public Service of Oklahoma, with subsequent changes in both utility rates and town finances.

Sebring was victimized by a 1979 decision to invest in an innovative diesel generating plant that proved to be both high cost and unnecessary to meet demand. Despite huge rate increases during the 1980s, the utility remained financially nonviable. Thereafter, the city began seeking a buyer, and, in 1993, sold the distressed plant to Tampa Electric and the local distribution network to Florida Power. By the time of the final sale, local residents were pleading to be rid of the entire public system.

These last examples reveal some of the limitations of public ownership. Most publicly owned electric utilities are small, often lacking in good financial controls or strong management, and vulnerable to reversals of fortune. In those cases, any alternative may appear preferable to the current state of affairs.

Systematic analysis of the effects of municipalization or privatization has proved impossible.[19] As illustrated here, most utilities' experiences are the product of unique factors in their operating environment. For that reason, it is difficult to offer confident generalizations.

7.4 Summary

This chapter is directed at the question of the determinants of public versus private ownership. While incomplete, the analysis has identified certain important causal forces. Some, like utility size, have been anticipated by earlier results. Others, like electricity consumption, are predicted by theories of interest group pursuit of objectives. Beyond these, the magnitude of utilities' contributions and services and popular election of state utility commissioners have both been found to be plausibly related to economic forces.

Apart from the inherent interest in these questions, the results have some potential implications for the statistical analysis. To the degree that public ownership not only affects certain variables but also may be affected by them, the econometric assumption that the variable *PUBLIC* is exogenous may be subject to reexamination. Unfortunately, a perfect disentangling of the effects of public ownership would require a yet more complex model and yet more data, the latter a particularly daunting task.

Notes

1. These theories and much related evidence are reviewed in Noll (1989). More recently, see, for example, Kaserman, Mayo, and Pacey (1993).

2. While their example is the London Stock Exhange, Hart and Moore's analysis is quite general.

3. Skew is defined as a larger difference between mean and median characteristics. As Hart and Moore note, skew is a stronger condition than heterogeneity. A uniform distribution would represent heterogeneity, but not skew.

4. A separate literature has examined the impact of voting over the choice of income tax schedules. See, in particular, Roberts (1977) and Wright (1986).

5. In a related vein, Lopez-de-Silanes, Shleifer, and Vishny (1995) examine the non-performance determinants of the choice of public versus private ownership, that is, "clean government laws," laws restricting county spending, and strong public unions.

6. The key difference among localities involves the preferred proportion of total cost of service to be paid by riders versus local taxpayers, a choice reflecting local demographic characteristics such as residents' age, income, motor vehicle registration, and so forth.

7. Square miles of service territory or miles of local distribution network are not reliably reported by all utilities. DENSITY represents the best effort at a consistent series (based on distribution miles), but since it cannot be computed for a considerable number of observations, those must be omitted in certain regressions.

8. Of course, it has been apparent from the outset that publicly owned utilities serve significantly smaller numbers of customers than do IOUs.

9. OLS regression analysis is used, since the dependent variable *ELECTED* is not limited to values of zero and one. For some utilities that span state boundaries, *ELECTED* is calculated as their weighted average values and thus takes on other values. Excluding them and conducting a probit analysis makes little difference.

10. Somewhat larger are payments in lieu of taxes ("tax equivalents"), which have been discussed previously. Their magnitude is often (but not always) governed by formula. Here we focus on discretionary items — ad hoc contributions and services rendered.

11. Regional dummy variables are also included throughout.

12. The most important local tax offset by the utility is property taxes, whose incidence is skewed towards upper income residents.

13. Meter count and customer count are approximately equal.

14. Vince (1990) (quoted by Lesch in the appendix to Coopers & Lybrand, 1995).

15. Coopers & Lybrand (1993) offers a good discussion of these issues.

16. The key legal issue would appear to be whether this represents a "sham" wholesale transaction intended to bring about *retail* wheeling. Since FERC is specifically prohibited from imposing the latter, such a determination would undermine "muni lite." See Allen and Allen (1995); also, *Petition of Virginia Electric and Power Co.* v. *City of Falls Church, VA, for Declaratory Judgment*, filed March 13, 1995.

17. This experience and those described next are taken from Coopers & Lybrand (1993, 1995) and Schweitzer (1995).

18. In a virtually identical scenario in Brook Park, Ohio, Ford supported an inquiry into municipalization, ultimately secured eight million dollars in rate concessions from Cleveland Electric Illuminating, and then shared part of its savings with Brook Park, which refunds $185 per year to each resident utility customer.

19. These accounts are taken from Woolbright (1990) and Weaver (1993).

20. Despite considerable efforts during the course of this study, reliable price and cost data on utilities that have undergone conversions could not be compiled.

Chapter 8

CONCLUSIONS

The purpose of this study has been to analyze the principal dimensions of performance of the U.S. electric power industry, and to identify and quantify their structural determinants. In so doing, this study has described this important industry in detail, with particular attention to the roles of public versus private ownership, integration versus deintegrated operation, and competition versus regulated monopoly. Most especially, we have been interested in how these and other forces actually determine industry costs and prices.

Though not an explicit part of the study, the evidence also casts light on the likely effects of proposals for transforming the industry. The reason is that most reform proposals and many reforms already underway involve changes in ownership, integration, and competition — the very characteristics of the industry examined here.

8.1 Ownership, Integration, and Competition

This research has several noteworthy features relative to previous studies of the electric power industry. First, it utilizes an extremely comprehensive database rather than modest numbers of select utilities. This study encompasses almost all the major public and private utilities covering nearly 90 percent of electric power sales. Possible concerns over selection problems or other sample bias are thus substantially alleviated if not altogether eliminated.

Second, both the cost and price of electric power are specified as integral dimensions of utility performance, rather than either in isolation. Among other advantages, such a perspective allows the relationship of cost *to* price to be explored, and in so doing, disentangles the effect of various factors on cost from their effect on price. This, too, helps to avoid errors in identifying causal forces.

A third significant feature of the present research is that the major determinants of utility performance are examined jointly, rather than focusing on each separately. Inclusion of multiple determinants in the same framework addresses concern that some omitted factor might be responsible for apparent effects.

Fourth, a considerable effort has been made to utilize theory in formulating models for empirical test. Instead of the ad hoc approach, characteristic of a great deal of past research, theoretically sound models of cost functions, markup pricing, segment effects, and so forth are developed. This addresses the concern that empirical findings might be the result of somewhat arbitrary specifications.

And lastly, this research extends each area of inquiry in new and informative directions. A fully-simultaneous equations model of pricing, cost, and demand is developed. Differences in scale and vertical economies between public and private utilities are allowed for. Pricing is analyzed by customer segment. The role of regulator selection method is examined. All of these represent important advances in the analysis of industry performance.

A wide range of substantive questions has been covered in this study. This range reflects not only the inherent importance of such questions, but also the enormous diversity of the U.S. electric power industry that permits such an inquiry. Public versus private ownership, vertical integration versus deintegrated operation, and distribution monopoly versus competition are the primary differences represented in the industry and examined in this study. Substantial evidence has been developed on the relationship of each difference to utility price and cost performance. While no short summary can do justice to this multifaceted study, some highlights are worth noting.

With respect to ownership, less than 10 percent of electric utilities in the U.S. are the familiar investor-owned companies. Out of about 3,000 electric utilities, most in fact are publicly owned. The latter are typically smaller, more closely tied to their communities, and often simply distribute power. Given their franchised monopoly status, IOUs are subject to state regulation. These contrasting features of IOU and publicly owned utilities afford a direct comparison of alternative techniques for social control.

While theory and empirical research have previously examined public ownership and regulation, there has been no agreement about their relative performance. The data and methodology in this study, however, clearly establish that publicly owned utilities have lower costs than comparable IOUs — 5.5 percent lower overall. Moreover, their lower costs appear to arise in the distribution function, attesting to the "comparative advantage" of public systems in end-user tasks. By contrast, investor-owned utilities achieve greater efficiency in generation where localism is less important and in terms of vertical relationship between generation and distribution.

These results provide a striking cost-based explanation for the observations that most publicly owned utilities are pure distributors and that most power is generated by investor-owned utilities. In addition, the disparity between the type of utility that achieves superior performance in distribution versus generation suggests that it is not regulation that lies at the root of the phenomenon, since both functions are subject to similar regulation. Rather, it would appear to be the distinctive advantages of public ownership and control of the end-user distribution service.

Further analysis indicates that public systems also price their power on average lower than do IOUs. The difference is estimated at between 2.5 percent and 3.7 percent in alternative models. Disaggregation into customer classes reveals that this overall average reduction is comprised of a very large reduction in residential price from public ownership (approximately 15 percent), a more modest decrease in commercial price (nearly five percent), and a small *increase* in industrial users' price (three percent). Collectively, these results (as well as others from an analysis of the incidence of public ownership and of popular election of state utility commissioners) support the proposition that public ownership of electric utilities is often the result of pressures from residential customers, who are, in turn, its major beneficiaries.

A second significant difference among utilities that has been addressed at length in this study concerns the degree of vertical integration. The complete spectrum of integration between generation and the downstream transmission/distribution functions is represented among U.S. electric utilities, from pure distribution utilities to those that generate all that they distribute. The high degree of vertical integration in that portion of the industry accounting for most power supply, however, has long suggested important vertical economies, although actual evidence on this issue has been very sparse.

This study has found that electric power production is characterized by significant vertical economies between generation and distribution, as well as significant economies of scale for each. The average utility in the data set enjoys a 22 percent cost saving from integrated production of its output amounts. This result provides a cost rationale for the dominant industry structure observed, namely, integrated generation and distribution. For utilities that may not be thoroughly integrated, holding company membership appears to be a cost-saving alternative, whereas power pools on average are not associated with lower costs.

A striking disparity emerges between the vertical economies achieved by public versus private utilities. At their respective average volumes, IOUs realize a substantial cost savings from integration, while publicly owned systems enjoy considerably smaller benefits. Together with the analysis of overall or multistage economies, these results also help to explain the fact that most large and privately owned utilities are in fact highly integrated while publicly owned utilities are specialists in distributing purchased power rather than in generation.

A third significant area of interest concerns competition among utilities at the retail or distribution stage. Some form of retail competition occurs in a number of jurisdictions where a second distributor competes for current or new customers, or operates side by side in a fashion that facilitates cost or rate comparisons. Although these cases have been noted before, difficulties in identifying the nature of "competition" and in measuring possible effects have left the importance of the issue in some doubt.

This study substantially resolves that ambiguity. Actual competition among utilities is found to reduce both costs and price. The overall price effect of about 2.5 percent is not always statistically significant. It is considerably larger in situations where current customers can switch between competing utilities (about eight percent), and larger yet for the publicly owned utility than its rival IOU (10.6 percent versus 1.5 percent). Nonetheless, it is clear that competition in some form has considerable potential in electric power markets to bring about efficiency benefits in both costs and prices.

Thus, the present study has addressed key questions about electric utility performance, utilizing data and modeling that represent significant improvements over previous research. It has clarified and perhaps even resolved several longstanding issues about the role of ownership, integration, and competition on utility performance. And strikingly, it has also demonstrated the economic bases for the diversity of the industry. The coexistence of public and private utilities, their differences in vertical integration, and the distribution orientation of publicly owned utilities are all shown to follow from underlying cost considerations. The frequency and pattern of public ownership, as well as popular election of utility commissioners, are part of a process by which residential customers in particular strive to realize pricing benefits.

Although some of these results have been anticipated in the literature, this study has established their validity, identified relevant mechanisms, shown their interrelationships, and quantified the phenomena. In other respects, this study goes well beyond others. Notably, important distinctions have been drawn between the cost effects of public ownership in distribution versus generation, and among the price effects of varying degrees of distribution competition. As we shall also see, the results are useful in considering industry reforms.

8.2 Restructuring the Electric Power Industry

The electric power industry is on the agenda for reform in many countries. Some countries, in fact, have already begun that process, by replacing public ownership with private enterprise, full integration of generation, transmission and distribution with deintegration, and traditional monopoly regulation with significant elements of competition. But most countries have approached reform of structure and governance of electric utilities with considerable caution, more so than transformations of other industries such as airlines and telecommunications. The reason is the widespread belief that electric power represents a different and more difficult set of circumstances. Network economies; close interdependence between stages of production; persistent monopoly power, at least in some portions of the industry; and the essential character of electric power to consumers — all of these factors increase the risk and the cost of policy errors.

It has not been the purpose of this study to evaluate the numerous proposals for transforming the electric power sector. But this research has significant implications for that process, since it has analyzed the impact of the very issues on which most reforms center — namely, privatization, deintegration, and competition. Here we shall set out, with appropriate caveats, some of those implications for reform.

Over the past 20 years, privatization has become a standard prescription for improving the performance of enterprises worldwide. The motivation varies in detail but is almost always rooted in the belief that public ownership results in cost inefficiency, arbitrary pricing, and misuse of the enterprise for political purposes. Present evidence demonstrates that this belief is far too simplistic and in some respects, simply incorrect..

Publicly owned enterprises are not necessarily inefficient nor is their pricing uniformly above some norm. Rather, it appears that publicly owned utilities that specialize in end-user tasks such as distribution may well have advantages over their privately owned counterparts. The most likely reason for this appears to be that retail distribution — of electricity and perhaps other goods and services — may be performed better by enterprises closely rooted to the customer community. Such proximity may yield greater knowledge of local customer needs and a greater sense of responsibility for addressing those needs.

A second possible interpretation of the evidence, not necessarily inconsistent with the above, is that public enterprise may have an advantage over private ownership where the latter is subject to regulation. Since rate-of-return regulation has been the norm for privately owned electric utilities in the U.S., present evidence is conditional upon regulation. The results of this study indicate that regulation may well be responsible for suboptimal performance but that the relative performance of publicly owned and regulated privately owned utilities differs by function. That suggests, in turn, that regulation is not uniformly the cause of a performance difference. Some other factor that differs by function must be responsible.

The third possible explanation noted at the outset of this study involves various subsidies to publicly owned utilities. The evidence clearly confirms the existence of such subsidies, but the very same evidence also demonstrates a significant cost difference between public and private utilities after all such subsidies have been accounted for. Public ownership, in short, achieves lower costs for reasons that go significantly beyond advantages due to subsidies.

The present conclusion may therefore apply to a wide range of policy choices: Simply shifting an enterprise to regulated private status will not necessarily lead to improvements in its performance. The task of improving efficiency requires careful attention to the tasks to be performed and to the governance mechanism of the enterprise.

One potentially very helpful tool in this task is competition. The merits of competition at the generation stage of electric power are well understood, but the role of competition at other stages has generally been thought to be limited by the cost disadvantages of reduced scale and/or facilities duplication. Nothing here implies that such concerns are unfounded. Yet the role of competition should not be underestimated. Even if *potential* costs are lower under monopoly provision, *realized* monopoly costs may be greater as a result of inefficiencies tolerated under regulation or public ownership. It is these latter effects that appear to be reduced by the forces of competition between utilities. This conclusion is tempered by the fact that observed competition arises in the context of existing regulation and public ownership, so that its effect is, strictly speaking, incremental. On the other hand, it seems reasonable to suppose that existing competition is constrained by regulatory mechanisms, so that the effect of competition in a less constrained environment might well be greater.

It should also be stressed that these observations about competition should not necessarily be taken too literally as a recommendation for duopoly distribution. The evidence is probably better interpreted as demonstrating the considerable potential for competition *in some form* to constrain costs and prices even in segments of the electric power industry with seemingly incompatible technology. Two such possibilities are retail wheeling and incentive regulation.

By allowing end users to shop for the lowest cost supply and requiring local distribution utilities to transmit and distribute that power, retail wheeling has the potential to introduce competition into an additional stage of the process. As a result, the local wire/distribution business might become quite distinct from a competitive "supply" business that actually arranges for electricity supply. An alternative form that competition might take entails market-oriented governance of public enterprise or incentive regulation. Despite the equivocal evidence found in this study, the potential for market-oriented, performance-based regulation to improve enterprise performance appears significant.

Of course, the potential for competition among generating entities has long been recognized and indeed represents a critical element in most reform proposals. At the same time, however, the most thoroughgoing technique for unleashing market forces — namely, deintegration of generation from distribution — has always raised concerns about possible sacrifice of vertical economies between stages of production. With two exceptions, this study confirms the existence of such economies and in fact finds them to be quite large.

One exception to this generalization involves small utilities, which actually encounter *dis*economies from integration and therefore appear appropriately to adopt stand-alone operation, typically as distributors. Publicly owned systems represent the second exception. The much smaller advantages of integration that they realize are consistent with — indeed, are presumably the reason for — their usual role as largely or purely distribution entities. Clearly, there is justification for diversity in the U.S. electric power industry rather than uniformity of firm sizes, structures, and ownership. The niche for each type of enterprise is ultimately determined by its comparative advantage.

These results in any event do not imply that vertical integration is mistaken policy for the electric power industry. Rather, they suggest that any viable proposal for restructuring the sector must provide for some alternative mechanism for realizing the economies that now derive from formal integration of generation and distribution. The failure to do so will lead to substantial cost inefficiencies, very likely large enough to offset any cost and price benefits from competition among generating companies.

Most proposals for industry restructuring do in fact address the relationship of generators to downstream distribution systems, and the differences among them often hinge precisely on the mechanism proposed for mediating this critical vertical relationship. Some advocate independent system operators of the grid, while others propose new types of power pools. Our purpose here is not to evaluate these alternatives, but rather to make clear that the stakes are high. Some mechanism capable of achieving the very substantial efficiencies presently derived from formal integration must exist or be developed if generation is to be severed from downstream stages of production. There is, in fact, some evidence that membership in holding companies can produce cost savings comparable to those that vertical integration otherwise provides. This and other alternatives require careful attention.

Deintegration, competition, and privatization are all policies very much in favor for electric power and industries generally. Reforms structured on these principles clearly can foster reductions in costs of production and in prices to customers, and so the present movement is by no means misguided. But the evidence demonstrates equally clearly the importance of vertical integration and of public ownership. For this reason, simple prescriptions for the industry — that it be privatized, that it be deintegrated, that it be deregulated, and so forth — too readily overlook the fact that the industry supports a diversity of forms, each more or less in the place where it has a comparative advantage and best contributes to common performance objectives.

Nothing in this study suggests that certain types of restructuring and reforms would not improve the performance of this industry. Deintegration is well suited to allowing competitive forces to operate in the generation stage and to narrow the scope of very imperfect regulation. Indeed, many have argued that without significant deintegration, efforts to create a more competitive environment are unlikely to succeed. But vertical restructuring must also be accompanied by a plan to achieve vertical economies in some other fashion. Absent the alternative, some large adverse effects may emerge.

Privatization is more of an issue in other countries than in the U.S., of course, but the present evidence still carries some implications. Sweeping plans to privatize all utilities appear too simplistic in that they overlook potential advantages of public ownership for distribution of power to end users. On the other hand, privatization of large utilities, fully integrated into generation, is more likely to result in cost savings.

Throughout, it is clear that the role of competition is critical. Competition imposes cost and price discipline on both privately owned, regulated utilities as well as on those that are publicly owned. It relieves the administrative burden on either of those mechanisms of enterprise governance, while simultaneously enhancing their effectiveness (or even superseding them) as a disciplining force. It provides the standard by which all enterprises' performance should be judged.

The challenge for policy, then, is to introduce elements of competition wherever possible into the evolving structure and operation of the electric power industry. That may well do more to ensure the success of reform than any other strategy. And it will allow for the continued viability of whatever range or diversity of vertical integration and ownership modes is appropriate for the electric power industry.

APPENDIX A
Data Definitions and Sources

Over 150 pieces of primary data on each of 543 utilities were compiled for this study. This Appendix sets out the definitions and sources of all the specific variables actually employed. For convenience of exposition, these are organized into broad categories. For reference purposes, a table of variables in alphabetic order appears as Appendix Table A.1. Appendix Table A.2 lists all the utilities in the database, together with their ownership, integration, and competition status.

Outputs, Purchases, and Integration

Distribution output is denoted *DIST* and defined as total megawatt-hours sold by each utility to final customers and to the resale market. Energy furnished without charge, that used internally by the utility, and line losses are excluded from *DIST*. For some purposes, distribution sales to final customers are disaggregated into residential, commercial, and industrial segments. Resale mwh (*RESALE)* is included as an exogenous variable in demand functions and in the disaggregate models.

Generation output (*GEN)* is simply total megawatt-hours generated by the utility. In the case of joint ventures, joint action agencies, and the like, allocations are made based on the best information available concerning ownership shares. Both distribution output and generation output also appear in squared forms, *DISTSQ* and *GENSQ*, and in the interaction form *DIST•GEN*.

Utility purchases of wholesale power (*PURCH*) are measured as total megawatt-hours.

All quantity data are taken from Form 861 (DOE, 1989a).

Prices and Costs

Average price (*PRICE*) is obtained by dividing total revenue by *DIST* and is measured in cents per kwh. Where appropriate, prices for residential sales (*PRRES*), commercial sales (*PRCOM*), and industrial sales (*PRIND*) are determined by dividing revenue in each customer class by class output. All revenue data are taken from Form 861.

Total cost is measured by total electric operating expenses. It is the sum of operations and maintenance expenses, depreciation and amortization, and an imputed return on invested capital. O & M for electric utilities are taken directly from DOE reports. Depreciation and amortization and interest on long-term debt are reported for the entire utility. For utilities with gas operations, these items are prorated according to the percent of their net capital devoted to electric power. The return on invested capital is obtained by multiplying net electric plant by the estimated cost of capital (see below).

Average cost (*AVGCOST*) is defined as total cost divided by distribution output *DIST*.

Sources of revenue and cost data are Forms 1 and 412 (DOE, 1989b, 1989c).

Ownership and Competition

Publicly owned utilities are identified by the fixed-effects variable *PUBLIC*. The interaction between public ownership and distribution output is denoted *PUB•DIST*, that with *GEN* is *PUB•GEN,* that with their interaction, *PUB•D•G*, and that with *PURCH* is *PUB•PURCH.*

Competition is measured in several alternative ways. *DUOPOLY* indicates utilities having duplicative distribution facilities where retail customers can switch. *HOOKUP* denotes those with duplicative distribution facilities in which customers can choose suppliers only upon initial hookup. *BORDER* represents cases where the boundary of two utilities' distribution networks pass through a municipality but their service territories do not overlap. Utilities falling into each of these categories are listed in text Tables 3.1, 3.2, and 3.3.

COMP designates all utilities falling into the *DUOPOLY, HOOKUP*, or *BORDER* categories, that is, those with *any* actual or potential form of "competition." The interactions of *DIST* with *COMP* and with *DUOP* are denoted *DIST•COMP* and *DIST•DUOP*, respectively. That between public ownership and duopolies is given by *PUB•DUOP*.

Other Affiliations and Institutions

Subsidiaries of registered holding companies are identified from Financial Statistics (DOE, 1989a) and denoted *HCSUB*.

Utilities that are members of regional power pools are denoted *POOLMEM*. Sources of information include Joskow and Schmalensee (1984), DOE (1981), and direct contact with DOE and utility personnel.

INCENTR designates IOUs that are subject to state incentive regulation, rather than traditional rate-of-return regulation. These cases are identified from compilations in Joskow and Schmalensee (1987) and NERA (1990).

GASUTIL denotes electric utilities that are diversified into natural gas distribution as well. DOE Forms 1 and 412 indicate utilities with assets and operations in gas distribution.

Interactions between distribution output and these variables are denoted *DIST•HC, DIST•POOL,* and *DIST•INCENT.*

Generation Capacity

For each utility that generates at least some power, the variable *FCGEN* takes on a value of unity. In addition, the percentages of its capacity comprised of nuclear, hydro, and non-steam "other" are defined and denoted as *NUCLEAR, HYDRO,* and *OTHER* respectively. The suppressed category is steam generation.

Data for IOUs are from Form 1, and for publicly owned utilities, Form 412.

Customer Characteristics

The number of customers in the residential, commercial, and industrial segments of the market are represented by *RESCUST, COMCUST,* and *INDCUST,* respectively. Their total is denoted *CUSTOM.* The percentage of residential customers relative to the total is *PCTRES.* Data are from Form 861.

The typical size of each type of customer — residential, commercial, and industrial — is captured by its average electric power usage. This is total mwh sales to a customer class divided by the number of customers in the class, denoted *RESSIZE, COMSIZE,* and *INDSIZE* for the three classes.

The percentage of each utility's output sold at high voltage is calculated as industrial output plus resale output, divided by total distribution output *DIST.* This is designated by *HIVOLT.*

For a subset of observations, number of residential customers per distribution cable mile can be calculated. This is denoted *DENSITY.*

Input Costs

Steam fuel cost per kwh generated by steam (*PRFSTE*) and nuclear fuel cost per nuclear-generated kwh (*PRFNUC*) are calculated by dividing each fuel cost total by the amount of power generated by that mode. These variables apply only to those utilities actually utilizing the input. Fuel cost data are from Forms 1 and 412 (DOE, 1989b, 1989c).

Since actual payroll data are available for only 367 utilities, the alternative employed is city or state average manufacturing wage (*WAGE*) for the appropriate area. For utilities operating in more than one state, a weighted average of the state wage data is created. The data source is Bureau of the Census (1991).

Capital Cost

The calculation of the cost of capital variable (*COSTCAP*) begins with the familiar weighted average cost of common stock, preferred stock, and long-term debt (Atkinson and Halvorsen; Hayashi et al., Henderson, and Pescatrice and Trapani). The cost of common stock is taken as the fourth-quarter (1989) dividends paid, annualized and divided by year-end stock price. The cost of preferred stock is total preferred dividends paid, divided by the book value of preferred stock. The cost of long-term debt is defined as interest on long-term debt divided by long-term debt outstanding. For publicly owned utilities, which do not issue stock, the cost of capital reduces to their cost of long-term debt.

As discussed in the text, *COSTCAP* adjusts the above calculation to reflect two distinctive items in the capital structure of publicly owned systems. "Investment by municipality" and "constructive surplus/deficit" are arguably interest-free loans and are here treated as such.

Sources of data are Forms 1 and 412 (DOE, 1989b, 1989c), plus Duff and Phelps (1990).

Taxes and Other Payments

Tax rates must be defined somewhat differently for each type of utility. Direct taxes paid by IOUs are the sum of federal income taxes, other income taxes, and "other" taxes, together with an adjustment to account for the following: Privately owned utilities can avoid income tax to the extent that current-year credits from accelerated depreciation exceed previously deferred credits come due. This has typically been the case, but for those IOUs for which the reverse is true, their negative net balance increases current year tax liability and is therefore included as a tax payment.

Publicly owned systems pay direct taxes consisting of certain non-income taxes and tax equivalents. The composite of these payments by IOUs and publicly owned systems results in a first approximation of the tax rate per mwh for each utility.

The above aggregations of "direct" tax payments are augmented to reflect certain tax substitutes affecting publicly owned utilities. The latter usually make "contributions" (also known as "payments in lieu of taxes") and supply services of various kinds to their municipalities effectively in place of taxes, and municipalities in turn may provide some of the same to the utility. *TAXRATE* therefore includes these "contributions and services" by publicly owned utilities, net of any contributions and services provided by their governments to the utilities.

Data are from Forms 1 and 412 (DOE, 1989b, 1989c).

Regions

Dummy variables are defined for the nine regions set out by the National Energy Reliability Council.

Election

ELECTED denotes IOUs operating in states where public utility commissioners are elected by popular vote. Those in two or more states have sales-weighted average values of *ELECTED* for their constituent states.

Data are from NARUC (1990).

Demand Shifters

Disposable income per capita for each utility's service territory (*INCOMPC*) is approximated by the state average disposable income per capita. This is a weighted average for multistate utilities. Per capita income data are from the Bureau of the Census (1991).

The price of natural gas, the principal substitute fuel, is calculated for all three types of customers and denoted by *PRGASR*, *PRGASC*, and *PRGASI*. Data are from DOE (1990b).

Cooling degree days (*COOLDAY*) and heating degree days (*HEATDAY*) in each utility's area are taken from NOAA (1990, 1991).

Table A.1. Principal Variables Used and Their Definitions	
VARIABLE	**DEFINITION**
AVGCOST	Average cost per kwh sold.
BORDER	Fixed effects variable for a utility bordering another within a municipality.
COMCUST	Number of commercial customers.
COMP	Fixed effects variable for a hookup, duopoly, or border competition.
COMSIZE	Average kwh sales to commercial customers.
COOLDAY	Cooling degree days.
COSTCAP	Cost of capital.
CUSTOM	Total number of customers.
DENSITY	Residential customers per distribution mile.
DIST	Total megawatt-hours sold to final customers and to resale.
DISTSQ	DIST squared.
DUOPOLY	Fixed effects variable for a utility facing a fully competitive system.
ELECTED	Fixed effects variable for a utility that is regulated by elected utility commissioners.
FCGEN	Fixed effects variable for a utility that generates power.
GASUTIL	Fixed effects variable for a utility diversified into gas distribution.
GEN	Megawatt-hours of generation.
GENSQ	GEN squared.
HCSUB	Fixed effects variable for a subsidiary of a registered holding company.
HEATDAY	Heating degree days.
HIVOLT	Percentage of sales at high voltage.
HOOKUP	Fixed effects variable for a utility facing new hookup competition.

Table A.1. Principal Variables Used and Their Definitions (Continued)	
VARIABLE	**DEFINITION**
HYDRO	Percent of generating capacity comprised of hydro.
INCENTR	Fixed effects variable for a utility operating under incentive regulation.
INCOMPC	Per capita income of residential customers.
INDCUST	Number of industrial customers.
INDSIZE	Average kwh sales to industrial customers.
NUCLEAR	Percent of generating capacity comprised of nuclear.
OTHER	Percent of generating capacity not comprised of steam, nuclear, or hydro.
PCTRES	Residential customers as a percentage of the total.
POOLMEM	Fixed effects variable for a member of a regional power pool.
PRCOM	Average price per kwh sold to commercial customers.
PRFNUC	Price of nuclear fuel.
PRFSTE	Price of steam fuel.
PRGASC	Price of natural gas to commercial customers.
PRGASI	Price of natural gas to industrial customers.
PRGASR	Price of natural gas to residential customers.
PRICE	Average price per kwh sold.
PRIND	Average price per kwh sold to industrial customers.
PRRES	Average price per kwh sold to residential customers.
PUBLIC	Fixed effects variable for a publicly owned utility.
PURCH	Megawatt-hours of purchased power.
RESALE	Megawatt-hours sold to the resale market.

Table A.1. Principal Variables Used and Their Definitions (Continued)	
VARIABLE	**DEFINITION**
RESCUST	Number of residential customers.
RESSIZE	Average kwh sales to residential customers.
TAXRATE	Tax rate.
WAGE	Wage rate of labor.

Table A.2. UTILITIES IN DATA SET

Utility Name	Ownership	Generation	Competition
Alabama			
Alabama Power Co	I	+	Y
Albertville City of	P	0	N
Andalusia City of	P	0	N
Athens City of	P	0	N
Bessemer City of	P	0	Y
Cullman Power Board	P	0	N
Decatur City of	P	0	N
Dothan City of	P	0	N
Florence City of	P	0	N
Foley City of (Riviera Utils)	P	0	N
Fort Payne Improvement Auth	P	0	N
Guntersville City of	P	0	N
Huntsville City of	P	0	N
Muscle Shoals City of	P	0	N
Opelika City of	P	0	N
Scottsboro City of	P	0	N
Sheffield City of	P	0	N
Sylacauga City of	P	0	N
Troy City of	P	0	N
Tuskegee City of	P	0	N
Arizona			
Arizona Public Service Co	I	+	Y
Citizens Utilities Co	I	+	N
Mesa City of	P	0	Y
Navajo Tribal Utility Auth	P	0	N
Salt River Proj Ag I & P Dist	P	+	Y
Tucson Electric Power Co	I	+	N
Arkansas			
Arkansas Power & Light Co	I	+	Y
Bentonville City of	P	0	N
Conway Corp	P		N
Hope City of	P	0	N
Jonesboro City of	P	+	N
North Little Rock City of	P	+	Y
Osceola City of	P	0	N
Paragould Light & Water	P	0	N
Siloam Springs City of	P	0	N

Table A.2. UTILITIES IN DATA SET

Utility Name	Ownership	Generation	Competition
West Memphis City of	P	+	N
California			
Alameda City of	P	0	N
Anaheim City of	P	+	N
Azusa City of	P	0	N
Burbank City of	P		N
Colton City of	P	0	Y
Glendale City of	P	+	N
Imperial Irrigation District	P	+	N
Lodi City of	P	0	N
Los Angeles City of	P	+	N
Modesto Irrigation District	P	+	N
Pacific Gas & Electric Co	I	+	Y
Palo Alto City of	P	0	N
Pasadena City of	P	+	N
Redding City of	P	0	N
Riverside City of	P	+	N
Roseville City of	P	0	N
Sacramento Municipal Util Dist	P	+	N
San Diego Gas & Electric Co	I	+	N
Santa Clara City of	P	+	N
Southern California Edison Co	I	+	Y
Turlock Irrigation District	P	+	N
Colorado			
Centel Corp	I	+	N
Colorado Springs City of	P	+	Y
Fort Collins City of	P	0	N
Longmont City of	P	+	N
Loveland City of	P	+	N
Public Service Co of Colorado	I	+	N
Connecticuit			
Connecticut Light & Power Co	I	+	Y
Groton City of	P	0	N
Norwich City of	P	+	N
United Illuminating Co	I	+	N
Wallingford Town of	P	+	N
Deleware			
Delmarva Power & Light Co	I		N
Dover City of	P	+	N

Table A.2. UTILITIES IN DATA SET

Utility Name	Ownership	Generation	Competition
Newark City of	P	0	N
District of Columbia			
Potomac Electric Power Co	I	+	Y
Florida			
Bartow City of	P	0	N
Florida Power & Light Co	I	+	Y
Florida Power Corp	I	+	N
Fort Pierce Utilities Auth	P	+	N
Gainesville Regional Utilities	P	+	N
Gulf Power Co	I	+	N
Homestead City of	P	+	N
Jacksonville Beach City of	P	0	N
Jacksonville Electric Auth	P	+	N
Key West City of	P	+	N
Kissimmee Utility Authority	P	+	N
Lake Worth City of	P	+	Y
Lakeland City of	P	+	N
Leesburg City of	P	+	N
New Smyrna Beach Utils	P	+	N
Ocala City of	P	+	N
Orlando Utilities	P	+	N
Quincy City of	P	0	N
Sebring Utilities	P	+	N
St Cloud City of	P	+	N
Tallahassee City of	P	+	N
Tampa Electric Co	I	+	N
Vero Beach City of	P	+	Y
Georgia			
Albany Water Gas & Light	P	0	Y
Calhoun City of	P	0	N
Cartersville City of	P	0	N
College Park City of	P	0	N
Covington City of	P	0	N
Crisp County Power	P	+	N
Dalton City of	P	+	N
Douglas City of	P	0	N
East Point City of	P	0	N
Fitzgerald Wtr Lgt & Bond	P	0	N
Georgia Power Co	I	+	Y

Table A.2. UTILITIES IN DATA SET

Utility Name	Ownership	Generation	Competition
Griffin City of	P	0	N
La Grange City of	P	0	N
Marietta City of	P	0	N
Moultrie City of	P	0	N
Savannah Electric & Power Co	I	+	N
Thomasville City of	P	0	N
Idaho			
Idaho Falls City of	P	+	N
Illinois			
Batavia City of	P	0	N
Central Illinois Light Co	I	+	N
Central Illinois Pub Serv Co	I	+	Y
Commonwealth Edison Co	I	+	N
llinois Power Co	I	+	Y
Iowa-Illinois Gas&Electric Co	I	+	N
Mt Carmel Public Utility Co	I	0	N
Naperville City of	P	0	N
Springfield City of	P	+	N
St Charles City of	P	0	N
Indiana			
Anderson City o f	P	0	N
Auburn City o f	P	0	N
Bluffton City o f	P	+	N
Crawfordsville Elec Lgt&Pwr Co	P	+	N
Frankfort City of	P	0	N
Indiana Michigan Power Co	I	+	N
Indianapolis Power & Light Co	I	+	N
Jasper City of	P	+	N
Logansport City of	P	+	N
Mishawaka City of	P	0	N
Northern Indiana Pub Serv Co	I	+	N
Peru City of	P	+	N
Public Service Co of IN Inc	I	+	N
Richmond City of	P	+	N
Southern Indiana Gas & Elec Co	I	+	N
Washington City of	P	0	N
Iowa			
Ames City of	P	+	N
Cedar Falls City of	P	+	N

Table A.2. UTILITIES IN DATA SET

Utility Name	Ownership	Generation	Competition
Interstate Power Co	I	+	N
Iowa Electric Light & Power Co	I	+	N
Iowa Power Inc	I	+	N
Iowa Public Service Co	I	+	N
Iowa Southern Utilities Co	I	+	N
Muscatine City of	P	+	N
Kansas			
Coffeyville City of	P	+	N
Garden City City of	P	0	N
Kansas City City of	P	+	N
Kansas Gas & Electric Co	I	+	N
Kansas Power & Light Co	I	+	Y
McPherson City of	P	+	N
Winfield City of	P	+	N
Kentucky			
Bowling Green City of	P	0	N
Frankfort City of	P	0	N
Franklin City of	P	0	N
Glasgow City of	P	0	N
Henderson City Utility	P	+	N
Hopkinsville City of	P	0	N
Kentucky Power Co	I	+	N
Kentucky Utilities Co	I	+	Y
Louisville Gas & Electric Co	I	+	N
Madisonville City of	P	0	N
Mayfield City of	P	0	N
Murray City of	P	0	N
Owensboro City of	P	+	N
Paducah City of	P	0	N
Union Light Heat & Power Co	I	0	N
Lousiana			
Alexandria City of	P	+	Y
Central Louisiana Elec Co Inc	I	+	Y
Lafayette City of	P	+	Y
Louisiana Power & Light Co	I	+	Y
Morgan City City of	P	+	N
Natchitoches City of	P	0	N
New Orleans Public Service Inc	I	+	Y
Ruston City of	P	+	N

Table A.2. UTILITIES IN DATA SET

Utility Name	Ownership	Generation	Competition
Terrebonne Parish Consol Gov't	P	+	Y
Maine			
Bangor Hydro-Electric Co	I	+	N
Central Maine Power Co	I	+	N
Maine Public Service Co	I	+	N
Maryland			
Baltimore Gas & Electric Co	I	+	Y
Easton Utilities	P	+	N
Hagerstown City of	P	0	N
Potomac Edison Co	I	+	N
Massachusetts			
Boston Edison Co	I	+	Y
Braintree Town of	P	+	N
Cambridge Electric Light Co	I	+	N
Chicopee City of	P	+	Y
Commonwealth Electric Co	I	+	N
Concord Town of	P	0	N
Danvers Town of	P	0	N
Eastern Edison Co	I	0	N
Fitchburg Gas & Elec Light Co	I	+	N
Hingham City of	P	0	N
Holyoke City of	P	+	N
Hudson Town of	P	+	N
Littleton Town of	P	+	N
Mansfield Town of	P	0	N
Massachusetts Electric Co	I	0	Y
Middleborough Town of	P	0	N
North Attleborough Town of	P	+	N
Norwood City of	P	0	N
Peabody City of	P	+	N
Reading Town of	P	0	N
Shrewsbury Town of	P	+	N
Taunton City of	P	+	N
Wakefield Town of	P	0	N
Wellesley Town of	P	0	N
Western Massachusetts Elec Co	I	+	Y
Westfield City of	P	0	N
Michigan			
Bay City City of	P	+	Y

Table A.2. UTILITIES IN DATA SET

Utility Name	Ownership	Generation	Competition
Coldwater Board of Public Util	P	+	N
Consumers Power Co	I	+	Y
Detroit Edison Co	I	+	N
Edison Sault Electric Co	I	+	N
Grand Haven City of	P	+	N
Holland City of	P	+	Y
Lansing City of	P	+	N
Marquette City of	P	+	N
Michigan Power Co	I	+	Y
Sturgis City of	P	+	N
Traverse City City of	P	+	Y
Upper Peninsula Power Co	I	+	N
Wyandotte Municipal Serv	P	+	N
Minnesota			
Alexandria City of	P	0	N
Anoka City of	P	0	Y
Austin City of	P	+	N
Fairmont Public Utilities	P	+	N
Hutchinson Utilities	P	+	N
Marshall City of	P	+	N
Minnesota Power & Light Co	I	+	N
Moorhead City of	P	0	N
New Ulm Public Utilities	P	+	N
Northern States Power Co	I	+	Y
Otter Tail Power Co	I	+	N
Owatonna City of	P	0	N
Rochester Public Utilities	P	+	N
Willmar Municipal Utils	P	+	N
Mississsipi			
Aberdeen City of	P	0	N
Clarksdale City of	P	+	N
Columbus City of	P	0	N
Greenwood Utilities	P	+	N
Holly Springs City of	P	0	N
Louisville Electric System	P	0	N
Mississippi Power & Light Co	I	+	N
Mississippi Power Co	I	+	N
New Albany City of	P	0	N
Starkville City of	P	0	N

Table A.2. UTILITIES IN DATA SET

Utility Name	Ownership	Generation	Competition
Tupelo City of	P	0	N
West Point City of	P	0	N
Missouri			
Carthage City of	P	0	N
Columbia City of	P	+	N
Hannibal City of	P	0	N
Independence City of	P	+	N
Kansas City Power & Light Co	I	+	Y
Kirkwood City of	P	0	Y
Marshall City of	P	+	N
Poplar Bluff City of	P	+	Y
Rolla City of	P	0	N
Sikeston City of	P	+	Y
Springfield City of	P	+	N
St Joseph Light & Power Co	I	+	N
Union Electric Co	I	+	Y
UtiliCorp United Inc	I	+	Y
Montana			
Montana Power Co	I	+	N
Montana-Dakota Utilities Co	I	+	N
Nebraska			
Cornhusker Public Power Dist	P	0	N
Dawson County Public Pwr Dist	P	0	N
Fremont City of	P	+	N
Grand Island City of	P	+	N
Hastings City of	P	+	N
Lincoln Electric System	P	+	N
Loup River Public Power Dist	P	+	N
Nebraska Public Power District	P	+	N
Norris Public Power District	P	0	N
North Platte City of	P	0	N
Omaha Public Power District	P	+	N
Southern Nebraska Rural P P D	P	0	N
Nevada			
Nevada Power Co	I	+	N
Sierra Pacific Power Co	I	+	N
New Hampshire			
Public Service Co of NH	I	+	N

Table A.2. UTILITIES IN DATA SET

Utility Name	Ownership	Generation	Competition
New Jersey			
Atlantic City Electric Co	I	+	Y
Jersey Central Power&Light Co	I	+	Y
Public Service Electric&Gas Co	I	+	Y
Rockland Electric Co	I	0	Y
Vineland City of	P		Y
New Mexico			
Farmington City of	P	+	N
Gallup City of	P	0	N
Los Alamos County	P	+	N
Public Service Co of NM	I	+	N
Texas-New Mexico Power Co	I	+	N
New York			
Central Hudson Gas & Elec Corp	I	+	N
Consolidated Edison Co-NY Inc	I	+	Y
Fairport Village of	P	0	N
Freeport Village of Inc	P	+	N
Jamestown City of	P	+	Y
Long Island Lighting Co	I	+	Y
New York State Elec & Gas Corp	I	+	Y
Niagara Mohawk Power Corp	I	+	Y
Orange & Rockland Utils Inc	I	+	N
Plattsburgh City of	P	0	N
Rochester Gas & Electric Corp	I	+	N
Rockville Centre Village of	P	+	N
Solvay Village of	P	0	N
North Carolina			
Albemarle City of	P	0	N
Carolina Power & Light Co	I	+	Y
Concord City of	P	0	N
Duke Power Co	I	+	Y
Elizabeth City City of	P	+	N
Fayetteville Public Works	P	+	N
Gastonia City of	P	0	N
Greenville Utilities	P	0	N
High Point Town of	P	0	Y
Kinston City of	P	0	N
Lexington City of	P	0	N
Lumberton City of	P	0	N

Table A.2. UTILITIES IN DATA SET

Utility Name	Ownership	Generation	Competition
Monroe City of	P	0	N
Morganton City of	P	0	N
Murphy City of	P	0	N
Nantahala Power & Light Co	I	+	N
New Bern City of	P	0	N
New River Light & Power Co	P	0	N
Rocky Mount City of	P	0	N
Shelby City of	P	0	N
Statesville City of	P	0	N
Tarboro Town of	P	0	N
Washington City of	P	0	N
Wilson City of	P	0	N
Ohio			
Bowling Green City of	P	0	N
Bryan City of	P	+	N
Celina City of	P	0	N
Cincinnati Gas & Electric Co	I	+	N
Cleveland City of	P	+	Y
Cleveland Electric Illum Co	I	+	Y
Columbus City of	P	+	Y
Columbus Southern Power Co	I	+	Y
Cuyahoga Falls City of	P	0	N
Dayton Power & Light Co	I	+	Y
Dover City of	P	+	N
Hamilton City of	P	+	N
Niles City of	P	0	N
Ohio Edison Co	I	+	Y
Ohio Power Co	I	+	Y
Orrville City of	P	+	N
Painesville City of	P	+	N
Piqua City of	P	+	Y
Toledo Edison Co	I	+	N
Wadsworth City of	P	0	N
Westerville City of	P	0	Y
Oklahoma			
Altus City of	P	0	N
Claremore City of	P	0	N
Edmond City of	P	0	N
Empire District Electric Co	I	+	N

Table A.2. UTILITIES IN DATA SET

Utility Name	Ownership	Generation	Competition
Oklahoma Gas & Electric Co	I	+	N
Ponca City City of	P	+	N
Public Service Co of Oklahoma	I	+	Y
Stillwater Utilities Authority	P	+	N
Oregon			
Ashland City of	P	+	N
Central Lincoln Peoples Utl Dt	P	0	N
Clatskanie Peoples Util Dist	P	0	N
Columbia River Peoples Ut Dist	P	0	N
Emerald Peoples Utility Dist	P	0	N
Eugene City of	P	+	N
Forest Grove City of	P	0	N
Idaho Power Co	I	+	N
McMinnville City of	P	0	N
Northern Wasco County P U D	P	0	N
PacifiCorp	I	+	Y
Portland General Electric Co	I	+	Y
Springfield City of	P	0	N
Tillamook Peoples Utility Dist	P	0	N
Pennsylvania			
Chambersburg Borough of	P	+	N
Duquesne Light Co	I	+	Y
Metropolitan Edison Co	I	+	Y
Pennsylvania Electric Co	I	+	N
Pennsylvania Power & Light Co	I	+	Y
Pennsylvania Power Co	I	+	Y
Philadelphia Electric Co	I	+	N
UGI Corp	I	+	N
West Penn Power Co	I	+	Y
Rhode Island			
Blackstone Valley Electric Co	I	+	N
Narragansett Electric Co	I	+	N
Newport Electric Corp	I	+	N
South Carolina			
Camden City of	P	0	N
Easley Combined Utility System	P	0	N
Gaffney City of	P	0	N
Greenwood City of	P	0	N
Greer City of	P	0	Y

Table A.2. UTILITIES IN DATA SET

Utility Name	Ownership	Generation	Competition
Lockhart Power Co	I	+	N
Newberry City of	P	0	N
Orangeburg City of	P	+	N
Rock Hill City of	P	0	N
South Carolina Electric&Gas Co	I	+	N
South Carolina Pub Serv Auth	P	+	N
South Dakota			
Black Hills Corp	I	+	N
Brookings City of	P	0	N
Northwestern Public Service Co	I	+	N
Watertown City of	P	0	N
Tennessee			
Alcoa Utilities Utilities	P	0	N
Athens City of	P	0	N
Benton County	P	0	N
Bolivar City of	P	0	N
Bristol City of	P	0	N
Brownsville City of	P	0	N
Carroll County	P	0	N
Chattanooga City of	P	0	N
Clarksville City of	P	0	N
Cleveland City of	P	0	N
Clinton City of	P	0	N
Columbia City of	P	0	N
Cookeville City of	P	0	N
Covington City of	P	0	N
Dayton City of	P	0	N
Dickson City of	P	0	N
Dyersburg Electric System	P	0	N
Elizabethton City of	P	0	Y
Erwin Town of	P	0	N
Etowah City of	P	0	N
Fayetteville City of	P	0	N
Gallatin City of	P	0	N
Greeneville City of	P	0	N
Harriman City of	P	0	N
Humboldt City of	P	0	N
Jackson City of	P	0	N
Johnson City City of	P	0	Y

Table A.2. UTILITIES IN DATA SET

Utility Name	Ownership	Generation	Competition
Kingsport Power Co	I	0	N
Knoxville Utilities Board	P	0	N
LaFollette City of	P	0	N
Lawrenceburg City of	P	0	N
Lebanon City of	P	0	N
Lenoir City City of	P	0	N
Lewisburg City of	P	0	N
Lexington City of	P	0	N
Loudon City of	P	0	N
Maryville City of	P	0	N
McMinnville City of	P	0	N
Memphis City of	P	0	N
Milan City of	P	0	N
Morristown City of	P	0	N
Murfreesboro City of	P	0	N
Nashville City of	P	0	N
Newport City of	P	0	N
Oak Ridge City of	P	0	N
Pulaski City of	P	0	N
Ripley City of	P	0	N
Rockwood City of	P	0	N
Sevier County Electric System	P	0	N
Shelbyville City of	P	0	N
Sringfield City of	P	0	N
Sweetwater City of	P	0	N
Tullahoma City of	P	0	N
Union City City of	P	0	N
Weakley Municipal Electric Sys	P	0	N
Texas			
Austin City of	P	+	N
Brenham City of	P	0	N
Brownsville Public Utils Board	P	+	N
Bryan City of	P	+	N
Central Power & Light	I	+	N
College Station City of	P	0	N
Denton City of	P	+	N
El Paso Electric Co	I	+	N
Garland City of	P	+	Y
Georgetown City of	P	0	N

Table A.2. UTILITIES IN DATA SET

Utility Name	Ownership	Generation	Competition
Greenville City of	P	+	N
Gulf States Utilities Co	I	+	Y
Houston Lighting & Power Co	I	+	N
Kerrville Public Utility Board	P	0	N
Lubbock City of	P	+	Y
New Braunfels City of	P	0	N
San Antonio City of	P	+	N
Seguin City of	P	+	N
Southwestern Electric Power Co	I	+	N
Southwestern Electric Serv Co	I	0	N
Southwestern Public Service Co	I	+	Y
Texas Utilities Electric Co	I	+	Y
Weatherford Mun Utility System	P	+	N
West Texas Utilities Co	I		N
Utah			
Bountiful City City of	P	+	N
Logan City of	P	+	N
Murray City of	P	+	N
Provo City Corp	P	0	N
St George City of	P	+	N
Vermont			
Burlington City of	P	+	N
Central Vermont Pub Serv Corp	I	+	N
Green Mountain Power Corp	I	+	N
Virginia			
Bedford City of	P	+	N
Bristol Utilities Board	P	0	N
Danville City of	P	+	N
Harrisonburg City of	P	0	N
Manassas City of	P	+	N
Martinsville City of	P	+	N
Radford City of	P	+	N
Salem City of	P	0	N
Virginia Electric & Power Co	I		Y
Virginia Tech Electric Service	P	+	N
Washington			
Centralia City of	P	+	N
Ellensburg City of	P	0	N
PUD No 1 of Benton County	P	0	N

Table A.2. UTILITIES IN DATA SET

Utility Name	Ownership	Generation	Competition
PUD No 1 of Chelan County	P	+	N
PUD No 1 of Clallam County	P	0	N
PUD No 1 of Clark County	P	0	N
PUD No 1 of Cowlitz County	P	+	N
PUD No 1 of Douglas County	P	+	N
PUD No 1 of Franklin County	P	0	N
PUD No 1 of Grays Harbor Cnty	P	+	N
PUD No 1 of Klickitat County	P	0	N
PUD No 1 of Lewis County	P	+	N
PUD No 1 of Okanogan County	P	0	N
PUD No 1 of Pend Oreille Cnty	P	+	N
PUD No 1 of Snohomish County	P	+	N
PUD No 2 of Grant County	P	+	N
PUD No 2 of Pacific County	P	0	N
PUD No 3 of Mason County	P	0	N
Port Angeles City of	P	+	N
Puget Sound Power & Light Co	I	+	Y
Richland City of	P	0	N
Seattle City of	P	+	N
Tacoma City of	P	+	Y
Vera Irrigation District #15	P	0	Y
Washington Water Power Co	I	+	Y
West Virginia			
Appalachian Power Co	I	+	N
Monongahela Power Co	I	+	Y
Wheeling Power Co	I	0	N
Wyoming			
Gillette City of	P	0	N

Legend:

Ownership   - P denotes publicly owned
                 I denotes investor-owned
Generation - 0 denotes no generation (unintegrated)
                 + denotes some generation
Competition - Y denotes utility facing some competition
                 N denotes monopoly utility

APPENDIX B

APPENDIX B
Review of Cities with Competitive Utilities

These summaries are based on utility responses to the author's questionnaire[1] together with Reinemer (1987a, 1987b) and other sources noted below.

Bushnell, Illinois

The Bushnell Municipal Utility competes with Central Illinois Public Service for all customers within the city limits of Bushnell, Illinois. Nonprice competition characterizes efforts to win customers. Despite the relative ease of switching, relatively few customers actually do so in a typical year. The Municipal Utility purchases power through the Illinois Municipal Electric Agency and transmits it over Central Illinois Public Service lines.

Cleveland, Ohio

Cleveland represents the largest competitive jurisdiction in the country. Cleveland Public Power and Cleveland Electric Illuminating Co. engage in direct competition in about half of that city where CPP currently has facilities. Customers can switch suppliers fairly readily, and a significant number do so. Day-to-day competition is on a nonprice basis, although CPP claims a 30 percent rate advantage. The companies have one joint-use transmission line, but otherwise build and operate separate distribution facilities. They do not engage in purchase or sale of power with each other. Rather, CPP purchases most of its power from out-of-state utilities and contracts with CEI for transmission services.

Columbus, Ohio

Columbus Southern Power competes with the city-owned electric distribution service in parts of Columbus, Ohio. Customers are offered various incentives to switch, including free weatherization audits and senior citizens' discounts, among others. A modest number of customers are reported to make such a switch in any year. Both utilities sometimes use each other's poles, relying upon a "gentlemen's agreement" regarding attachments. In some places, the City also utilizes Columbus Southern's transmission lines.

Culpeper, Virginia

Virginia Power faces the town-owned electric utility in Culpepper, Virginia. While all customers can switch and the town even offers free tree trimming and other services as inducements, few customers choose to do so in any year. The utilities share poles, and the town also purchases transmission services from Virginia Power.

Dowagiac, Michigan

The City of Dowagiac, Michigan, competes with the Indiana & Michigan Power Co. in the city. Competition is on service rather than price, and entails advertising directly to customers. Switching requires a written release from the serving utility to confirm bill payment, plus the usual deposit. Few existing customers in fact switch. The utilities have a joint use agreement with respect to pole sharing. In addition, the city utility purchases power from I&M.

Duncan, Oklahoma

The City of Duncan and the Public Service Company of Oklahoma compete within the city limits. All customer classes can switch, and a significant number of customers do so. The city has adopted PSO's rates, but nets out the franchise fee and offers a discount for quick payment. Both utilities provide a variety of free and low-cost services. There are no jointly owned or used facilities.

Floydada, Texas

Southwestern Public Service competes with the city-owned electric company within the city limits of Floydada, Texas. Relatively easy switching results in a substantial number of customers who choose to do so. The city reports that it seeks new customers through advertising and unspecified free services, whereas SPS contends that "Neither City nor S.P.S. tries to get the others' customers." They share no facilities, but the city purchases all its power from SPS on a long-term contractual basis.

Lubbock, Texas

Southwestern Public Service also faces a city-owned distribution utility in Lubbock, Texas. Customers interested in switching must give a three-day advance notice, during which the serving utility is said to "have a chance to convince the customer to stay with them." Both companies are said to have "a new business department that tries to attract the competitor's customers daily," but competition is essentially on a non-price basis. To execute a change, a customer's meter will be switched and drop line reconnected to the new supplier. A significant number of customers switch suppliers in any year.

Lubbock Power & Light and Southwestern Public Service share no facilities. Rather, trunk lines are duplicated. The cost of drop lines supposedly is ignored. SPS and the local telephone company generally share one set of poles, while LP&L shares another set with the local cable operator (Bellamy, 1981). The city purchases about 40 percent of its power from SPS. Cooperation between the two electric companies is illustrated by the following practice (questionnaire response, City of Lubbock):

> During emergencies, customer outages, either company hooks
> the customer up to the other utility to reduce the time of the
> outage to the customer. When the utility having the problem has
> outage to the customer. When the utility having the problem
> has fixed the problem, then the customer is restored to that
> company.

Newton Falls, Ohio

The Newton Falls Utility Department and Ohio Edison compete within the city limits of Newton Falls, Ohio. A modest number of customers take advantage of the opportunity to switch suppliers. The city reports that its rates are 30 percent lower that Ohio Edison's. There is no sharing of facilities or purchase of power between the two.

Paris, Kentucky

City of Paris Combined Utilities competes with Kentucky Utilities in Paris. All customers may switch between the two, but only a small number do so. The city states that its rates are lower and that it provides faster service, but the utilities are prohibited from direct solicitation of each other's customers. A pole-sharing agreement between the city, Kentucky Utilities, and SouthCentral Bell provides for a two-dollar-per-pole charge for use of a pole belonging to any one of them. The two electric utilities also have a joint tree-trimming agreement. Paris Combined Utilities purchases most of its power requirements from KU.

Piqua, Ohio

Piqua Municipal Power and Dayton Power & Light compete in Piqua, Ohio. Despite fairly easy switching, few consumers in fact avail themselves of the opportunity, in part a reflection of the fact that Piqua Municipal Power now has most of the consumers. There is no advertising, price, or service competition and no shared facilities. Piqua Municipal Power does, however, purchase short-term power from Dayton Power & Light.

Traverse City, Michigan

Until August, 1994, Traverse City Light & Power competed with Consumers Power Co. in Traverse City, Michigan. In addition, Cherryland Rural Electric Cooperative also operated in part of the city. Switching was quite simple and a modest number of customers in fact switched suppliers. Price and service competition characterized the market. Use of poles was shared between Traverse City and Consumers Power (but not with Cherryland Electric) on a rental basis specified by contract. Traverse City also purchased transmission services from Consumers Power.

In late 1994, an agreement was signed under which the city utility is to provide service to 375 of Consumers Power's customers. Within ten years, Traverse City will purchase half of Consumers Power's distribution system and associated customers and, in an additional ten years, nearly all the remainder. In adjacent areas, customer switching is restricted and new customers generally must hook up with the closest supplier.[2] Competition has effectively ceased.

Notes

1. This questionnaire is described in the text.
2. "Customer-Switching Fight is Settled by Michigan City, Consumers Power," *Public Power Weekly*, August 22, 1994.

BIBLIOGRAPHY

Alchian, Armen, and Harold Demsetz, "Production, Information Costs, and Economic Organization," *American Economic Review*, December 1972.

Allen, Donald, "Municipalization 'Lite': A Quicker Way to Access Competitive Markets," paper presented at Infocast Conference on Competitive Power Sourcing for Industrial Companies, Chicago, March 1995.

American Public Power Association, *Survey of Administrative and Policymaking Organization of Publicly Owned Utilities: 1995*. Washington, DC: 1995.

American Public Power Association, "Public Power Sellouts (1980-94)" and "Publicly Owned Utilities Established (1980-94)," Washington, DC, undated.

American Public Power Association, *Payments and Contributions by Public Power Distribution Systems to State and Local Government: 1992*. Washington, DC: July 1994.

Anderson, Douglas, *Regulatory Politics and Electric Utilities: A Case Study in Political Economy*. Auburn House: Boston, 1981.

Atkinson, Scott, and Clifford Nowell, "Explaining Regulatory Commission Behavior in the Electric Utility Industry," *Southern Economic Journal*, January 1994.

Atkinson, Scott and Robert Halvorsen, "The Relative Efficiency of Public and Private Firms in a Regulated Environment: The Case of U.S. Electric Utilities," *Journal of Public Economics*, April 1986.

Atkinson, Scott and Robert Halvorsen, "A Test of Relative and Absolute Price Efficiency in Regulated Utilities," *Review of Economics and Statistics*, February 1980.

Averch, Harvey and Leland Johnson, "Behavior of the Firm Under Regulatory Constraint," *American Economic Review*, December 1962.

Bailey, Elizabeth, and Ann Friedlaender, "Market Structure and Multiproduct Industries," *Journal of Economic Literature*, September 1982.

Baker, Jonathan, "Identifying Cartel Policing Under Uncertainty: The U.S. Steel Industry," *Journal of Law and Economics*, October 1989.

Baron, David, "Design of Regulatory Mechanisms and Institutions," in *Handbook of Industrial Organization*, edited by R. Schmalensee and R. Willig. Amsterdam: North-Holland, 1989.

Barrington-Wellesley Group, *A Diagnostic Audit of the Los Angeles Department of Water and Power*, May 1994.

Baumol, William, John Panzar, and Robert Willig, *Contestable Markets and the Theory of Industry Structure*. New York: Harcourt, Brace, and Jovanovich, 1982.

Becker, Gary, "A Theory of Competition Among Pressure Groups for Political Influence," *Quarterly Journal of Economics*, August 1983.

Beierlein, James, James Dunn, and James McConnon, "The Demand for Electricity and Natural Gas in the Northeastern United States," *Review of Economics and Statistics*, August 1981.

Berg, Sanford, and Jinook Jeong, "An Evaluation of Incentive Regulation for Electric Utilties," *Journal of Regulatory Economics*, March 1991.

Berg, Sanford, and John Tschirhart, *Natural Monopoly Regulation*. Cambridge: Cambridge University Press, 1988.

Bernard, Jean-Thomas, and Michel Roland, "Rent Dissipation Through Electricity Prices of Publicly Owned Utilities," Department of Economics, Universite Laval, Quebec, April 1995.

Berndt, Ernst, *The Practice of Econometrics*. Reading, MA: Addison-Wesley, 1991.

Berndt, Ernst, and David Wood, "Technology, Prices, and the Derived Demand for Energy," *Review of Economics and Statistics*, August 1975.

Berry, Dan, "Private Ownership Form and Productive Efficiency: Electric

Cooperatives vs. Investor-Owned Utilties," *Journal of Regulatory Economics*, December 1994.

Bierce, Ambrose, *The Devil's Dictionary*. Sagamore, New York 1957.

Boardman, Anthony and Aidan Vining, "Ownership and Performance in Competitive Environments: A Comparison of the Performance of Private, Mixed, and State-owned Enterprises," *Journal of Law and Economics*, April 1989

Bohi, Douglas, *Analyzing Demand Behavior: A Study of Energy Elasticities*. Baltimore: Johns Hopkins Press, 1981.

Bohn, Roger, Michael Caramanis, and Fred Schweppe, "Optimal Pricing in Electric Networks over Space and Time," *Rand Journal of Economics*, Autumn 1984.

Boyer, Kenneth, "Can Market Power Really Be Estimated?" *Review of Industrial Organization,* February 1996.

Bresnahan, Timothy, "Empirical Studies of Industries with Market Power," in *Handbook of Industrial Organization*, edited by R. Schmalensee and R. Willig. Amsterdam: North-Holland, 1989.

Bresnahan, Timothy, "The Oligopoly Solution Concept Is Identified," *Economics Letters*, 1982.

Bureau of the Census, *Statistical Abstract of the United States*, Washington, 1991

Carlton, Dennis, and Jeffrey Perloff, *Modern Industrial Organization*. New York: HarperCollins, 1995.

Caves, Douglas and Laurits Christensen, "The Relative Efficiency of Public and Private Firms in a Competitive Environment: The Case of Canadian Railways," *Journal of Political Economy*, October 1980.

Christensen, Laurits, and William Greene, "An Econometric Assessment of Cost Savings from Coordination in U.S. Electric Power Generation," *Land Economics*, May 1978.

Christensen, Laurits, and William Greene, "Economies of Scale in Electric Power Generation," *Journal of Political Economy*, August 1976.

180

Coopers & Lybrand, *Electricity Municipalization Review 1995*, April 1995.

Coopers & Lybrand, *Electricity Municipalization and the Energy Policy Act of 1992*, February 1993.

Corazzini, Arthur, "Tax-Exempt Municipal Bond Finance: The Electric Utility Issue." APPA: Washington, DC, April 1989

Cowing, Thomas and V. Kerry Smith, "The Estimation of Production Technology: A Survey of Econometric Analyses of Steam-Electric Generation," *Land Economics*, May 1978.

Cramer, Curtis, and John Tschirhart, "Power Pooling: An Exercise in Industrial Coordination," *Land Economics*, February 1983.

DeAlessi, Louis, "An Economic Analysis of Government Ownership and Regulation," *Public Choice*, Fall 1974.

DeAlessi, Louis, "Some Effects of Ownership on the Wholesale Prices of Electric Power," *Economic Inquiry*, December 1975.

Department of Energy, *Financial Statistics of Selected Electric Utilities, 1989*, Washington, 1990a.

Department of Energy, *Natural Gas Annual 1989*, Energy Information Administration, Washington, 1990b.

Department of Energy, *Inventory of Power Plants in the United States 1993*. Washington, Energy Information Adminsitration, 1994.

Department of Energy, *Power Pooling in the U.S.*, Washington, 1981.

Department of Energy, *Typical Electric Bills*. Washington, Energy Information Administration, 1988.

Department of Energy, *Annual Electric Utility Report* (Form EIA-861), Energy Information Administration, 1989a.

Department of Energy, *Annual Report of Major Electric Utilities* (FERC Form 1), Federal Energy Regulatory Commission, 1989b.

Department of Energy, *Annual Report of Public Electric Utilities* (Form EIA-412), Energy Information Administration, 1989c.

Department of Energy, *Electric Sales, Revenues, and Bills*, Energy Information Agency, 1989d.

Department of Energy, "Electricity Transmission: Realities, Theory and Policy Alternatives," Federal Energy Regulatory Commission, October 1989e.

DiLorenzo, Thomas and Ralph Robinson, "Managerial Objectives Subject to Political Market Constraints: Electric Utilities in the U.S.," *Quarterly Review of Economics and Business*, Summer 1982.

Doyle, Chris, and Maria Maher, "Common Carriage and the Pricing of Electricity Transmission," *Energy Journal*, 1992.

Duff & Phelps, *Public Utility Stock Stats*, Chicago 1990.

Einhorn, Michael, *From Regulation to Competition: New Frontiers in Electricity Markets*. Boston: Kluwer, 1994.

Ely, Richard T., *Outlines of Economics*. Hunt and Easton: New York, 1984.

Emmons, William, "Franklin D, Roosevelt, Electric Utilities, and the Power of Competition," *Journal of Economic History*, December 1993.

Fare, R., S Grosskopf, and J. Logan, "The Relative Performance of Publicly-Owned and Privately-Owned Electric Utilities," *Journal of Public Economics*, February 1985.

Geddes, Richard, "A Historical Perspective on Electric Utility Regulation," *Regulation*, Winter 1992.

Gegax, D. and John Tschirhart, "An Analysis of Interfirm Cooperation: Theory and Evidence from Electric Power Pools," *Southern Economic Journal*, April 1984.

General Accounting Office, *Electricity Supply: The Effects of Competitive Power Purchases Are Not Yet Certain*, Washington, DC, August 1990.

Gilsdorf, Keith, "Vertical Integration Efficiencies and Electric Utilities: A Cost

Complementarity Perspective," *Quarterly Review of Economics and Finance*, Fall 1994.

Gilsdorf, Keith, "Testing for Subadditivity of Vertically Integrated Electric Utilities," *Southern Economic Journal*, July 1995.

Gordon, Richard, "The Public Utility Holding Company Act," *Regulation*, Winter 1992.

Halvorsen, Robert, "Residential Demand for Electric Energy," *Review of Economics and Statistics*, February 1975.

Hansmann, Henry, "Ownership of the Firm," *Journal of Law, Economics, and Organization*, Fall 1988.

Harris, Malcolm, and Peter Navarro, "Does Electing Public Utility Commissioners Bring Lower Electric Rates?" *Public Utilities Fortnightly*, September 1983.

Hart, Oliver, and John Moore, "The Governance of Exchanges: Members' Cooperatives vs. Outside Ownership," mimeo, Harvard University, April 1994.

Hausman, William and John Neufeld, "Property Rights versus Public Spirit: Ownership and Efficiency of U.S. Electric Utilities Prior to Rate-of-Return Regulation," *Review of Economics and Statistics*, August 1991.

Hausman, William and John Neufeld, "Public versus Private Electric Utilities in the United States: A Century of Debate over Comparative Economic Efficiency," *Annals of Public and Cooperative Economics*, 1994.

Hayashi, Paul, Melanie Sevier, and John Trapani, "Pricing Efficiency under Rate-of-Return Regulation: Some Empirical Evidence for the Electric Utility Industry," *Southern Economic Journal*, January 1985.

Hayashi, Paul, Melanie Sevier, and John Tranpani, "An Analysis of Pricing and Production Efficiency of Electric Utilities by Mode of Ownership," in *Regulating Utilities in an Era of Deregulation*, edited by M. Crew. New York, 1987.

Hellman, Richard, *Government Monopoly in the Electric Utility Industries: A Theoretical and Empirical Study*. New York: Praeger, 1972.

Henderson, Stephen, "Cost Estimation for Vertically Integrated Firms: The Case of Electricity," in *Analyzing the Impact of Regulatory Change in Public Utilities*, M. Crew, ed. Lexington: Lexington Books, 1985.

Hollas, Daniel, Stanley Stansell, and E. Tylor Claggett, "Ownership Form and Rate Structure: An Examination of Cooperative and Municipal Electric Distribution Utilities," *Southern Economic Journal*, October 1994.

Hollas, Daniel, and Stanley Stansell, "Regulation, Ownership Form, and the Economic Efficiency of Rural Electric Distribution Cooperatives," *Review of Regional Studies*, Summer 1991.

Hollas, Daniel, and Stanley Stansell, "An Examination of the Effect of Ownership Form on Price Efficiency: Proprietary, Cooperative, and Municipal Electric Utilties," *Southern Economic Journal*, October 1988.

Houston, Douglas, "User-Ownership of Electric Transmission Grids," *Regulation*, Winter 1992.

Huettner, David and John Landon, "Electric Utilities: Scale Economies and Diseconomies," *Southern Economic Journal*, April 1978.

Hughes, Thomas. *Networks of Power*. Baltimore, MD: Johns Hopkins University Press, 1983.

Jarrell, Gregg, "The Demand for State Regulation of the Electric Utility Industry," *Journal of Law & Economics*, October 1978.

Jarrell, Gregg, "The Demand for Electric Utility Regulation," in *Electric Power: Deregulation and the Public Interest*, John Moorhouse, ed. Pacific Research Institute: San Francisco, 1986.

Joskow, Paul, "The Determination of the Allowed Rate of Return in a Formal Regulatory Hearing," *Bell Journal of Economics*, Autumn 1982.

Joskow, Paul and Richard Schmalensee, *Markets for Power*. Cambridge: MIT Press, 1985.

Joskow, P. and R. Schmalensee, "Incentive Regulation for Electric Utilities," *Yale Journal on Regulation*, Fall 1986.

184

Joskow, Paul, "Regulatory Failure, Regulatory Reform, and Structural Change in the Electric Power Industry," *Brookings Papers on Microeconomic Activity: 1989.* Washington: Brookings, 1989.

Joskow, Paul, "Expanding Competitive Opportunities in Electricity Generation," *Regulation*, Winter 1992.

Joskow, Paul, Nancy Rose, and Catherine Wolfram, "Political Constraints on Executive Compensation: Evidence from the Electric Utility Industry," *Rand Journal of Economics*, Spring 1996.

Kamerschen, David, and Herbert Thompson, "Nuclear and Fossil Fuel Steam Generation of Electricity: Differences and Similarities," *Southern Economic Journal*, July 1993.

Kaserman, David, and John Mayo, "The Measurement of Vertical Economies and the Efficient Structure of the Electric Utility Industry," *Journal of Industrial Economics*, September 1991.

Kaserman, David, John Mayo and Patricia Pacey, "The Political Economy of Deregulation: The Case of Intrastate Long Distance," *Journal of Regulatory Economics*, March 1993.

Koh, Dong Soo, Sanford Berg, and Lawrence Kenny, "A Comparison of Costs in Privately Owned and Publicly Owned Electric Utilities: The Role of Scale," *Land Economics*, February 1996.

Krautman, Anthony, and John Solow, "Economies of Scale in Nuclear Power Generation," *Southern Economic Journal*, July 1988.

Kwoka, John, "Price Squeezes in Electric Power: The New Battle of *Concord*," *Electricity Journal*, June 1992.

Laffont, Jean-Jacques, and Jean Tirole, "Privatization and Incentives," *Journal of Law, Economics, and Organization*, 1991.

Laffont, Jean-Jacques, "Industrial Policy and Politics," *International Journal of Industrial Organization,* February 1986.

Landon, John, "Theories of Vertical Integration and Their Application to the

Electric Utility Industry," *Antitrust Bulletin*, Spring 1983.

Laskin, Alan, "Accelerated Depreciation and the Investment Tax Credit: How Big Is the Subsidy for Investor-Owned Utilities?" Washington, DC: APPA, March 1988.

Leibenstein, Harvey, "Allocative Efficiency vs. 'X-Inefficiency," *American Economic Review*, June 1966.

Lopez-de-Silanes, Florencio, Andrei Sheifer, and Robert Vishny, "Privatization in the United States," Harvard Institute of Economic Research Discussion Paper 1723, May 1995.

Lynk, William, "Nonprofit Hospital Mergers and the Exercise of Market Power," *Journal of Law & Economics*, October 1995.

Mann, Patrick, and Edmond Seifried, "Pricing in the Case of Publicly Owned Electric Utilities," *Quarterly Review of Economics and Business*, Summer 1972.

Mayo, John, "Multiproduct Monopoly, Regulation, and Firm Costs," *Southern Economic Journal*, July 1984.

Meyer, Robert, "Publicly Owned vs. Privately Owned Utilities: A Policy Choice," *Review of Economics and Statistics*, November 1975.

Meyer, Robert, and Hayne Leland, "The Effectiveness of Price Regulation," *Review of Economics and Statistics*, November 1980.

Michaels, Robert, "Deregulating Electricity," *Regulation*, Winter 1992.

Moore, Thomas, "The Effectiveness of Regulation of Electric Utility Prices," *Southern Economic Journal*, April 1970.

National Economic Research Associates. *Incentive Regulation in the Electric Utility Industry*, 1990.

National Association of Regulatory Utility Commissioners, *Annual Report on Utility and Carrier Regulation*, 1990.

National Oceanic and Atmospheric Administration, *Historical Climatological*

186

Series 5-1, 5-2, National Climate Data Center, Asheville NC, 1990, 1991.

Nelson, Randy and Walter Primeaux, "The Effects of Competition on Transmission and Distribution Costs in the Municipal Electric Industry," *Land Economics*, November 1988.

Nelson, Randy, "The Effects of Competition on Publicly-Owned Firms," *International Journal of Industrial Organization*, April 1990.

Nelson, Jon, Mark Roberts and Emsley Tromp, "An Analysis of Ramsey Pricing in Electric Utilities," in *Regulating Utilities in an Era of Deregulation*, edited by M. Crew (New York: 1987).

Neuberg, Leland, "Two Issues in the Municipal Ownership of Electric Power Distribution Systems," *Bell Journal of Economics*, Winter 1977.

Noll, Roger, "The Politics of Regulation," in *The Handbook of Industrial Orgamization*," R. Schmalensee and R. Willig, eds. Noth-Holland, 1989.

Owen, Bruce, and Mark Frankena, *Electric Utility Mergers: Principles of Antitrust Analysis*. New York: Praeger, 1994.

Pechman, Carl, *Regulating Power: The Economics of Electricity in the Information Age*. Norwood: Kluwer, 1993.

Peltzman, Sam, "Pricing in Public and Private Enterprises: Electric Utilities in the United States," *Journal of Law and Economics*, April 1971.

Peltzman, Sam, "Towards a More General Theory of Regulation," *Journal of Law & Economics*, 1976.

Pescatrice, Donn and John Trapani, "The Performance and Objectives of Public and Private Utilities Operating in the United States," *Journal of Public Economics*, August 1980.

Peters, Lon, "Non-Profit and For-Profit Firms Electric Utilities in the United States: Pricing and Efficiency," *Annals of Public and Cooperative Economics*, Oct. – Dec. 1993.

Pollitt, Michael, *Ownership and Performance in Electric Utilities*. Oxford: Oxford

University Press, 1995.

Priest, George, "The Origins of Utility Regulation and the 'Theories of Regulation' Debate," *Journal of Law & Economics*, April 1993.

Primeaux, Walter, "An Assessment of X-Efficiency Gained Through Competition," *Review of Economics and Statistics*, February 1977.

Primeaux, Walter, "Estimate of the Price Effect of Competition: The Case of Electricity," *Resources and Energy*, 1985.

Primeaux, Walter, *Direct Electric Utility Competition: The Natural Monopoly Myth*, 1986.

Primeaux, Walter, and Randy Nelson, "An Examination of Price Discrimination and Internal Subsidization by Electric Utilities," *Southern Economic Journal*, July 1980.

Primeaux, Walter, and Mann, Patrick, "Regulator Selection Methods and Electricity Prices," *Land Economics*, February 1986.

Putnam, Hayes, and Bartlett, *Analysis of the Differences Among Alternative Forms of Utility Ownership in the United States*. Washington, DC: Edison Electric Institute, 1985.

Putnam, Hayes, and Bartlett, *Subsidies and Unfair Competitive Advantages Available to Publicly-Owned and Cooperative Utilities*. Washington, DC: Edison Electric Institute, 1994.

Reinemer, Vic, "Door-to-Door Competition," *Public Power*, May-June 1987.

Roberts, Kevin, "Voting Over Income Tax Schedules," *Journal of Public Economics*, December 1977.

Roberts, Mark, "Economies of Density and Size in the Production and Delivery of Electric Power," *Land Economics*, November 1986.

Roller, Lars-Hendrik, "Proper Quadratic Cost Functions with an Application to the Bell System," *Review of Economics and Statistics*, May 1990.

188

Salpukas, Agis, "The Rebellion in 'Pole City'," *New York Times*, October 10, 1995, p. D1.

Schmidt, Klaus, "The Costs and Benefits of Privatization: An Incomplete Contract Analysis," University of Bonn, January 1993.

Schmidt, Stephen, "Interest Groups, Cost Distribution, and Influence over Mass Transit Policy," Department of Economics Series #116, Union College, October 1995.

Schweitzer, Martin, "Municipal Electric Utilities: Establishment and Transformation," Oak Ridge National Laboratory, March 1995.

Shapiro, Carl and R. Willig, "Economic Rationales for the Scope of Privatization," Princeton University, 1990.

Shleifer, Andrei and R. Vishny, "Politicians and Firms," *Quarterly Journal of Economics*, November 1994.

Sing, Merrile, "Are Combination Gas and Electric Utilities Multiproduct Natural Monopolies?" *Review of Economics and Statistics*, August 1987.

St. Marie, Stephen, "Effectiveness of Incentive Regulation: Nuclear Power Plant Performance and Operating and Maintenance Costs," PhD Dissertation, George Washington University, January 1996.

Stigler, George, "The Theory of Regulation," *Bell Journal of Economics*, Spring 1971.

Stigler, George, and Claire Friedland, "What Can Regulators Regulate? The Case of Electricity," *Journal of Law and Economics*, April 1962.

Taylor, John, "The Demand for Electricity: A Survey," *Bell Journal of Economics*, Spring 1975.

Tennessee Valley Authority, *Report on FY 1989 Payments Under Section 13 of the TVA Act.* September 1989.

Ulm, David, "Analysis of the Rate Benefit from Preferential Access to Federal Hydro Power," APPA: Washington, DC, 1993.

Vince, Clinton, "Lessons Learned by Cities Looking at the Public Power Option," paper presented at the APPA Conference, Hilton Head, October 1990.

Viscusi, Kip, John Vernon, and John Harrington, *Economics of Regulation and Antitrust*. Cambridge: MIT Press, 1995.

Weaver, Danialle, "Sebring's Demise," *Public Power*, Sept. – Oct. 1993.

Woolbright, Shane, "Diary of a Sellout," *Public Power*, Jan. – Feb. 1990.

Wright, Randall, "The Redistributive Roles of Unemployment Insurance and the Dynamics of Voting," *Journal of Public Economics*, December 1986.

Yunker, J., "Economic Performance of Public and Private Enterprise: The Case of U.S. Electric Utilities," *Journal of Economics and Business*, Fall 1975.

Zardhoohi, Asghar, "Competition in the Production of Electricity," in *Electric Power: Deregulation and the Public Interest*, John Moorhouse, ed. San Francisco: Pacific Institute, 1986.

Name Index

Subject Index

195